Reflections about Contemporary Management

New Horizons in Management Sciences
Edited by Lukasz Sulkowski

Scientific Board:
Prof. Dr. German Chavez
Prof. Dr. Marcela Rebeca Contreras Loera
Prof. Dr. hab. Natalia Czuchraj
Prof. Dr. Geoff Goldman
Prof. Dr. hab. Barbara Kozuch
Prof. Dr. Claude Martin
Prof. Dr. Thomas P. Massey
Prof. Dr. hab. Bogdan Nogalski
Prof. Dr. hab. Roman Patora
Prof. Dr. Americo Salvidar
Prof. Dr. hab. Kazimierz Zimniewicz

Volume 7

Barbara Kożuch / Łukasz Sułkowski (eds.)

Reflections about Contemporary Management

Bibliographic Information published by the Deutsche Nationalbibliothek
The Deutsche Nationalbibliothek lists this publication in the Deutsche Nationalbibliografie; detailed bibliographic data is available in the internet at http://dnb.d-nb.de.

Library of Congress Cataloging-in-Publication Data
A CIP catalog record for this book has been applied for at the Library of Congress

This publication was supported by the Institute of Public Affairs of the Jagiellonian University.

ISSN 2194-153X
ISBN 978-3-631-71835-3 (Print)
E-ISBN 978-3-631-73619-7 (E-PDF)
E-ISBN 978-3-631-73620-3 (EPUB)
E-ISBN 978-3-631-73621-0 (MOBI)
DOI 10.3726/b11992

© Peter Lang GmbH
Internationaler Verlag der Wissenschaften
Berlin 2018
All rights reserved.

Peter Lang – Berlin · Bern · Bruxelles · New York · Oxford · Warszawa · Wien

All parts of this publication are protected by copyright. Any utilisation outside the strict limits of the copyright law, without the permission of the publisher, is forbidden and liable to prosecution. This applies in particular to reproductions, translations, microfilming, and storage and processing in electronic retrieval systems.

This publication has been peer reviewed.

www.peterlang.com

Contents

Preface 7

Introduction 9

Part I. The Achievements of Polish Praxeological School
in Management Sciences 11

Professor Bogusław Nierenberg, PhD The Jagiellonian University
Professor Witold Kieżun, PhD – the illustrious Polish praxeologist 13

Professor Alojzy Czech, PhD
Jan Zieleniewski (1901–1973) – Cracow period, early works 25

Part II. Reflective Management Dilemmas 35

Professor Łukasz Sułkowski, PhD
Metaparadigmatic perspective in the theory
of organizations and management sciences 37

Professor Barbara Kożuch, PhD & Mateusz Lewandowski, PhD
Contemporary organizational humanism – overview 53

Part III. Contemporary Management Dilemmas 69

Colonel Marek Bodziany, PhD & Paweł Kocoń, PhD
Military science and management science – methodological
connection in the context of culture of organization 71

Sławomir Olko, PhD
Management of the multilateral ventures in the networks and clusters in
creative industries from the perspective of the activity
network theory – methodological aspects 83

Anna Kaczorowska, PhD
Reflections on the effectiveness of the evaluation phase in public projects 97

Paweł Romaniuk
Trends of changes in risk management supported by an internal audit
in all bodies of public administration along with their evaluation 117

Katarzyna Szara, PhD
Creative capital and capabilities of its measurement
within an organization 127

Anna Pawłowska, PhD
Lifelong learning motivation at the age of 45+.
Some sources of limitations 139

Joanna Radomska, PhD
Risk associated with employee participation in the process of strategy
implementation versus company size 151

Marta R. Jabłońska, PhD
Consumer needs and implementation of new technologies in the
information society illustrated with the example of electric
vehicle market 161

Monika Stachowiak-Kudła, PhD
Evaluation as an instrument of higher education quality assurance 177

Marzena Papiernik-Wojdera, PhD
Patterns of growth rates and real sales growth in Polish industrial
companies in 2006–2012 189

Preface

The conference titled *Between Cognition and Method – Dilemmas of Reflective Management*, the outcome of which is this volume, was organized in Cracow on 8 and 9 May 2014 by the Institute of Public Affairs of the Jagiellonian University. This meeting of the representatives of management science had been preceded directly by a solemn meeting of the Senate of the Jagiellonian University, summoned on 10 May 2014 on the 650th Jubilee of the University. During the ceremony, on the initiative of the Faculty of Management and Social Communication of the Jagiellonian University, the highest academic distinction, the title of Honoris Causa, was conferred upon Professor Witold Kieżun. It is also to him whom both the Conference in May 2014 and this post-conference volume are dedicated.

The simultaneous celebration of the 650th Jubilee of the Jagiellonian University and awarding the honorary degree of Doctor Honoris Causa to Professor Witold Kieżun was not a coincidence. The motto of the jubilee speech of the Rector of the Jagiellonian University, Professor Wojciech Nowak, was: "Proud of our tradition, we think about the past and look to the future." Furthermore, His Magnificence underlined the importance of the ideas of wisdom and freedom for the functioning of the University throughout the centuries, and the role the University has played in the history of Poland. It is not difficult to find a link between the cultivated tradition of the pursuit of truth and a concern for the common good, recurring frequently on the occasion of the 650th Jubilee of the University founded by King Casimir the Great. This relation has also marked the attitude and work of Professor Witold Kieżun during his whole fascinating life.

The heroic participation of Professor Witold Kieżun in the Warsaw Uprising and his sturdy attitude against the Soviet totalitarianism in post-war years, proves his patriotism, which gained its beautiful uncompromising nature and firm opinions through the tradition of Polish Romanticism. His exceptional prowess on the battlefield – honoured with the Cross of Valour and the War Order of *Virtuti Militari* – gives a special meaning to the Professor's words about current Polish reality. Looking to the future was strikingly reflected in the engagement of Professor Witold Kieżun in the science of organization and management. First of all, it is characterized by linking theory and practice, as shown by his expert achievements in Poland, Burkina Faso, Canada and the USA, as well as by his way of managing a UN project of public administration reform in some Central African countries (Burundi, Rwanda). Secondly, the Professor's academic contribution is marked by a very original line of research activity. All those who know the works of Professor Witold Kieżun pay attention to his approach of linking his interest in public management and public-social partnership to the praxeological management school developed by his masters – Professor Tadeusz Kotarbiński and Professor Janusz Zieleniewski.

In this context, Professor Witold Kieżun can be regarded as an uncompromising critic of the Polish transitions in the last decades. However, his criticism does not concern the quantitative but the qualitative aspect and stems from the previously mentioned spirit of praxeology which Tadeusz Kotarbiński defined as the science of doing things the right way or – as he put it in the title of his famous treatise – the science of a good job.

The insight and clarity of Professor Kieżun's judgement was already visible in his excellent habilitation dissertation, *Autonomization of Organizational Units. From The Pathology of Organization* (1971), which brought him international recognition. The work presents, in an uncompromising manner, the pathology of bureaucracy for which punishment is the aim of control, and control itself – the basic purpose of the existence of an organization, as tersely encapsulated by Professor Bogdan Nogalski in his review of the works of Professor Witold Kieżun, prepared in relation to the procedure of granting the honorary degree of Doctor Honoris Causa of the Jagiellonian University. Furthermore, the diagnosed process of autonomization, occurring in organizations and consisting in an imbalance between basic functions and the regulatory and auxiliary ones, leads to the diffusion of an organization, the essence of which is a change of purpose – it is a means of action which becomes its principal objective. Bearing the weaknesses of organizational though – concerning not only the reality of the Polish People's Republic or the Polish Third Republic but also the commonly recognized organizational thought – stems from the very same praxeological attitude and building the general theory of efficient management which would impact social practice and the common good.

It also has to be realized that although the thoughts of Professor Kieżun are sometimes used in public (if not political) discourse, their constant source is still the Citizen's care for the condition of his Homeland. For this reason, in our attempt to pay respect to the attitude and achievements of Professor Witold Kieżun, we hereby present the results of our conference.

Jacek Ostaszewski
Dean of the Faculty of Management and Social Communication

Introduction

The interest in organizations and management that has existed for over a hundred years is not shrinking at all, but its character is changing. Despite the substantial number of publications devoted to the management of enterprises, as well as public and non-governmental organizations, there is still a need to provide general knowledge and interpretation of more complex questions to different groups of people. The processes of adjusting oneself to contemporary conditions require improvement of institutional, functional and instrumental aspects of management.

By its nature, management science serves its application. Practitioners appreciate especially the part of this dynamically developing field which directly serves the development of their organization. For this reason, they are eager to learn about the secrets of the art of management, adapting the solutions proposed by the authors to the needs of their organization. They search for advice on how to develop unequivocally defined and routinized problems characteristic of the contemporary organization management.

However, ensuring real development of management science while creating conditions so that the created works are applied in management practice requires a certain level of thoughtfulness. This will bring a definite result, that is, the skill of solving multidimensional and unique management problems.

An increased mindfulness in the life of the man of the 21st century, postulated in social sciences, cannot be omitted in management processes. The role of mindfulness in management is not only to solve problems but also, and above all, to lead to conscious examination of the nature and substance of phenomena, deepening considerations on the conditions and consequences of the preached ideas, the opinions expressed, and the actions taken. These issues are among those poorly recognized in the world literature.

The presented collective monograph constitutes to some extent an attempt to fill this recognized gap. It consists of three parts.

The first part starts with a text on relations between praxeology and organization and management science, that is, remarks on the fringes of the ceremony of granting Professor Kieżun the title of Doctor Honoris Causa of the Jagiellonian University on 10 May 2014, presented at the conference *Between Cognition and Method – Dilemmas of Reflective Management*. The next chapter, devoted to Jan Zieleniewski, the most important forerunner of management science in Poland, and also the promoter and mentor of Professor Witold Kieżun, includes a reference to this part.

The second part is devoted to selected aspects of reflective exploration of the contemporary approach to management, and an attempt to answer the question of what contemporary organizational humanism is. The fourth chapter includes deliberations on the modelling of the key areas of the Positive Organizational Potential as a symptom of operationalization of the positive management paradigm. A similar concept applies to deliberations on the business model of a public organization, and in particular its contextual conditioning, as well as analysis of the possibility of applying an idiographic and nomothetic approach in the practice of evaluating the higher education units in Poland.

In the third part, the authors concentrate on the dilemmas of the contemporary organization management, mainly in Poland, but also in relation to the international works devoted to management science.

The book contains selected papers presented at the conference *Between Cognition and Method – Dilemmas of Reflective Management*. It is addressed to recipients from various countries and cultures. It allows to look at the works of Polish authors from the perspective of what is specific to the national literature on this subject, and what is universal, irrespective of the latitude, tradition or other similar factors.

Prof. dr hab. Barbara Kożuch
Prof dr hab. Łukasz Sułkowski

Part I
The Achievements of Polish Praxeological School in Management Sciences

Professor Bogusław Nierenberg, PhD
The Jagiellonian University

Professor Witold Kieżun, PhD – the illustrious Polish praxeologist

(remarks delivered on the occasion of the Doctor Honoris Causa Award Ceremeony at the Jagiellonian University, presented at the conference entitled: "Między poznaniem a metodą – dylematy refleksyjnego zarządzania")

Abstract: Professor Witold Kieżun was given the title of Doctor Honoris Causa of the Jagiellonian University in 2014. His life and science choices have always been clear. His words capture keen attention of scholars and students alike, they are heeded by theorists and managers. His main contribution in this field was his description of different pathologies of Polish bureaucracy: gigantomania, obsession with luxury, corruption, and arrogance of those in power.

Keywords: praxeology, Witold Kieżun, biography, life choices, management theory

Introduction

In his *Essays,* Michel de Montaigne distinguished three categories of the mind. He claimed that there was an Abecedarian ignorance preceding all knowledge. In turn, "in the average understandings and the middle sort of capacities, the error of opinion is begotten, whereas the higher and nobler souls, more solid and clear-sighted, make up another sort (…) by a long and religious investigation (…) have discovered the mysterious and divine secret (…) have arrived at that supreme degree with marvellous fruit and confirmation (…) and enjoy their victory with great spiritual consolation, humble acknowledgment of the divine favour, reformation of manners, and singular modesty" (Montaigne, 1952, p. 151).

Undeniably, Professor Witold Kieżun ranks among the highest class of the mind specified by Montaigne. Thus, appearing in the role of a promoter awarding the title of Doctor Honoris Causa of the Jagiellonian University to the Professor, entails not only an honour and privilege, but also some trepidation: How to avoid depicting such an illustrious figure with trivial platitudes? How can one describe so eminent a Personage? How to fit in but a few pages almost a century of noble and righteous life? These questions must be asked and answered by anyone who embarks on such an attempt, for all those who, if only perfunctorily, have become acquainted with the Professor's scientific work, as well as facts from his life, must bow their head with reverence. And so do I, extremely pleased that the Jagiellonian University decided to bestow its most honorary laurels onto this outstanding Scholar and righteous Man, our graduate.

The 20th century was not kind to our Homeland, plaguing it with calamities. And yet, as if to counterbalance them, it also produced outstanding figures, among whom Professor Witold Kieżun ranks prominently. It is said that the role of individuals should not be overestimated. And yet, in difficult times, it is them who act as signposts. Their work and unwavering attitudes give weaker characters the hope that by joint efforts even the most formidable obstacles can be overcome.

Searching for words suitable to describe the Professor, I stumbled upon a review of his novel *Niezapomniane twarze* (*The Unforgotten Faces*). The review was written by the outstanding Polish poet, Zbigniew Herbert, the Professor's friend and companion from their university times. This is howe Herbert described the main character of the novel, Teofil, who seems to function as the Author's *alter ego*:

> "Teofil is someone of the Generation of Columbuses. He experienced all that was the share of those who had been born in those years: the war and occupation, the Warsaw Uprising, he served a prison term, he was sent to Gulag, he survived several years of thraldom under the Soviet rule. Ever searching for the truth, ever striving to be objective, he has encountered people of all sorts during his life. They were his colleagues, brothers-in-arms, superiors, subordinates, or simply acquaintances. Some of them became ingrained in his memory for good" (Kieżun 2008, p. 6).
>
> "Their story was put to paper by Teofil's friend, Professor Witold Kieżun, the Polish praxeologist lecturing on American universities and the author of numerous scholarly works, published both at home and abroad" (Kieżun 2008, p. 6).

These words from Zbigniew Herbert demonstrate that today, as Horace in the days of yore, Professor Kieżun might say: *Exegi monumentum aere perennius* (I have raised a monument more permanent than bronze). But he would never do that, he is too modest. I needed to quote Herbert's words in order to reach to the substance of the Professor's work. Although Herbert was not a scholar himself, he hit nail on the head when he described, in a lapidary manner, his friend as "the Polish praxeologist lecturing on American universities and the author of numerous scholarly works published both at home and abroad" (Kieżun, 2008, p. 6).

Person and creative output

In his book titles *Prakseologia. Wstęp do nauki o sprawnym działaniu* (*Praxeology. An Introduction to the Science of Efficient Action*), Tadeusz Kotarbiński, another eminent Polish scholar and Witold Kieżun's master and teacher,, invokes a thought from Wojciech Jastrzębowski: "to know the way, but not to follow it, is worse than straying through wilderness" (Kotarbiński, 1975, p. 489). To successfully perform this task, speaking in praxeological terms, one needs: the know-how, capability, vigour and fitness, but also, and above all else, respectability. This last characteristic, according to Kotarbiński, ought to be understood in the ethical sense, for every intellectual work requires a combination of the external world, that is, the earth and objects that belong to the earth, and the internal world, comprised of our vital forces. Professor Kieżun has always exhibited an exemplary attitude, even in the most direst of straits, even facing the loss of his life. It was so during the Warsaw Uprising, it was so in the Soviet Gulag, it was so in the period of communist dictatorship, and it remains so until this very day. Kieżun's last books (2011, 2012, 2013) present his, sometimes bitter if incisive, reflection concerning the last quarter of the 20th century and the transformations which were to affect our country. A person such as Professor Kieżun is fully entitled to point out the failings and errors we have made along the way. He never did so out of sheer discontent, but always out of his true emotion towards the Homeland, whose greater good and successful development were always close to his heart.

Professor Kieżun indicated various organizational errors, both small and great, which have affected our society. He wrote, for example, that within the period from 1990 to 1998, the Polish central administration had swollen in numbers from 46 thousand to 126.3 thousand employees. He pointed out that the example of a four-tier organizational structure in Warsaw's City Council is no doubt a peculiar negative world record (Kieżun, 2000).

His words capture the keen attention of scholars and students alike, they are heeded both by management theorists and managers. It is a pity that politicians do not follow their footsteps; if they did, they would chase all sorts of pathologies out of Poland, including: gigantomania, obsession with luxury, corruption, and the arrogance of those in power – with such an incisive understanding indicated and described by the Professor.

His academic works have always been met with a lively response. They have bedazzled with precision of reasoning and the beauty of his thought. Speaking of the Professor's scholarly output, one must state that it is indeed impressive. Comprising more than four hundred published works, it includes more than eighty two monographs, textbooks, and scripts. And yet, any attempt at a holistic evaluation of his output is doomed to failure since, in being so rich, so diverse, and having such outstanding cognitive, theoretical and methodological values, it is impossible to classify it under a single label. However, the most important feature in the Professor's academic work is his praxeological perspective on the issues he investigates. Thus, he is able to combine, at the highest possible level, practical experience with theoretical reflection. It must be also fully emphasised that, in his deliberations, Professor Kieżun never strayed off the course set by his moral compass because, as he insisted after his master Tadeusz Kotarbiński, praxeology is an efficiency of action understood as achieving the intended goal at minimal material and moral costs, but with observance of ethical principles, for "without this determinant, e.g. the camp in Auschwitz could be recognized as a model of efficiency" (Kieżun, 2011, p. 17).

Hence, if one chooses to view Professor Kieżun's academic work in this perspective, then it may be recognized as erected on praxeological foundations. Especially, if one takes into account one of the most simple, but at the same time most beautiful definitions of praxeology holding that it is a science of doing things the right way. Viewing his work, one may easily state that all he has accomplished has been pursued in the right way. It simply was a good scholarly work! As a disciple of Professor Tadeusz Kotarbiński and Professor Jan Zieleniewski, he took over the scholarly work of his masters to multiply it most notably. Let us invoke several facts which make it possible – although only in a cursory manner – to outline the Professor's person and scholarly output.

In 1949, Witold Kieżun started to work in the National Bank of Poland and tried to combine his duties with scientific activities. Initially, his interests were drawn to law and finances. Yet, it was the year 1961 that proved to be a watershed event in his career. It was then when he met Professor Jan Zieleniewski. Five years later, in the Main School of Planning and Statistics, he defended his doctoral dissertation entitled *Zarządzanie Oddziałem Banku Polskiego* (*Managing a Department of the National Bank of Poland*), promoted by Professor Zieleniewski. The fundamental thesis of the dissertation was the idea of humanization of labour. I put particular emphasis on the Professor's humanistic perspective, for the Faculty of Management and Social Communication of the Jagiellonian University is the only faculty in Poland with full academic rights in the field of management sciences in the humanities.

The 1960s saw a number of the Professor's publications – a fruit of his scientific research. During this time, Professor Kieżun suggested the concept of a director-intellectual (today we would rather say: an educated manager), who is aware of his ancillary role in the organization and concentrates upon creating its intellectual capital. It was also at this time when his works referring to these ideas came into existence. At the turn of the 1960s and 1970s, Professor Kieżun conducted research on autonomization, which later provided the foundation for his habilitation procedure. Under the notion of autonomization, Professor Kieżun described the phenomenon of pathology consisting in: replacing the main goal with a secondary objective or another main goal; changing the *modus operandi*, thereby engendering wastefulness; changing goals resulting in the means of action becoming the main goal; and disturbing the necessary balance between basic, regulatory, and auxiliary functions. The ideas thus formed were received with interest by the entire academic community of the period. A several-tens-of-pages long abstract with results of his research was successfully transferred to the West where it was translated into English, French, and Spanish. The 1970s saw the publication of the first edition of his academic textbook, which is still valued today, as is attested by its four editions. The following years saw little of his publishing activities on the Polish market. At that time, lecturing on Western universities, mainly in the USA and Canada, Professor Kieżun published mostly in English and French.

Despite reaching a retirement age, Professor Kieżun does not consider himself an old-age pensioner and he does not intend to succumb to this stereotype. Professor Kieżun was a promoter for fifteen Doctors of Philosophy and a host of Masters and Bachelors. Some of his students – for example, Krystyna Bolesta-Kukułka, Jolanta Szaban, Czesław Sikorski, and Bolesław Kuc – went on to become renowned Professors of economy and management. Today, they continue the scientific work of their mentor, just like he continued the toil of his masters – Professors Tadeusz Kotarbiński and Jan Zieleniewski.

However, his work cannot be contained within exclusively academic frames. Professor Kieżun is also an artist in the fullest meaning of this word: he writes poetry, composes, and engages in journalism. His output includes memoirs as well as journalistic books. Referring to his prose, Zbigniew Herbert writes that it is "saturated with all smells, flavours, and colours of the soil; despite being rather thin in volume, it makes an impression of a huge, apocalyptic mural from the walls of the Sistine Chapel of our nation" (Kieżun, 2008, p. 5).

Biography

Professor Kieżun was born on 6 February 1922 in Vilnius. Still before the outbreak of the WWII, his family moved from Vilnius to Warsaw. It was there where in 1939 he graduated from high school. This was also the year when the Second World War began. Teenage boys had to become adult men in a single moment. They went on to become the Generation of Columbuses, the generation for whom giving their life for their Homeland presented no dilemma. Professor Kieżun was part of this generation. He became involved in conspiracy, and when the time came to take up arms, he did not hesitate. In the Warsaw Uprising, he fought like few others of this time. He was a member of the *Harnaś* special task force where he fought in the *Gustaw* battalion. He adopted the pseudonym *Wypad* (*Sortie*). He took part in the successful assault on the Main Post Office, located at what is known today as the Warsaw Uprising Square, he participated in the capture of the Parish House at the Church of Holy Cross and Police Headquarters at Krakowskie Przedmieście. His bravery and commitment to the cause stirred admiration and won him a wide recognition.

The Warsaw Uprising Museum displays a big photograph of a young man in a battledress and helmet with a white-and-red band. The insurgent in the photograph is smiling. He had just seized a small arsenal, and the weapons he captured would reinforce the insurgents' firepower. He could have died in action during this mission, yet he survived. What is more, he captured fourteen prisoners-of-war, the same number of rifles, and two thousand rounds of ammo. And all this on his own. It came as no surprise that he was quickly promoted to the rank of second-lieutenant and General Tadeusz Bór-Komorowski awarded the twenty-years-old insurgent with the Cross of Valour. A couple of weeks later, Witold Kieżun's chest was decorated with the Silver Cross of the War Order of Virtuti Militari for new acts of valour.

Few survived the Uprising. Today, he is the only living member of the task force where he fought. He gives testimony to the truth of those times. Several thousand young people flocked to meet him in the Warsaw Uprising Museum. This bestows the struggle of these times with meaning.

Yet, anybody who thinks that the end of war had put an end to the suffering is gravely mistaken. Already in 1945, having escaped German captivity, Kieżun was arrested by the NKVD and confined to the prison cell at Montelupich in Kraków. During the investigation, he was spared no torment (including a mock execution), yet he did not betray a single person. On 23 May 1945, he was deported to the Soviet labour camp in the Karakum Desert, where he miraculously survived a tropical disease called beri-beri.

Upon his return to Poland, Witold Kieżun was once again sent to a labour camp, as the Communist authorities viewed him as suspicious. After the release, he decided to continue education in Kraków. He remembers the three-year-period he spent at the Jagiellonian University as a very important time of his life. In his autobiography (Kieżun, 2013, p. 279), he indicated three reasons of this importance: Zbigniew Herbert – a university friend; other illustrious characters of Kraków he had a chance to meet; and the friendly atmosphere of the Jagiellonian University.

During his student days at the Jagiellonian University, Professor Kieżun met Zbigniew Herbert, one of the most eminent poets in Poland's post-war history.

This is how he characterizes the beginnings of his friendship with Herbert in his memoires: "We were in the same class and I would frequently see him at the university. Soon, it turned out that we shared a similar passion for poetry and a youthful fascination for seeking the truth. He was very interested in my camp experience, especially my personal relation to the mortal danger. The Warsaw Uprising was a recurring topic of our conversations. At the time, I still had the youthful ease in expressing myself through the language of poetry, I wrote quite a few memoir-type poems from the time of my imprisonment. I treated this primitive work as a peculiar type of fun – quite similarly to Herbert. Not only did we share a critical view of the progressing sovietization of the country, but also a peculiar passion for philosophy" (Kieżun 2013, p. 279).

And here comes the second of the reasons for the esteem in which Professor Kieżun holds the Jagiellonian University – namely, the illustrious figures he had the chance of meeting during his university times in Kraków. For example, he attended, together with Herbert, lectures delivered by Professor Roman Ingarden. Both of them were fascinated with Ingarden's philosophy.

Thanks to his contacts with the circles associated with the "Tygodnik Powszechny" weekly, young Witold Kieżun met Jerzy Turowicz, Antoni Gołubiew, Paweł Jasienica, and Stanisław Stomma. In 1949, the Editor-in-Chief Turowicz invited a young student called Kieżun to a vacation facility for the clergy in Kuźnice, the so-called Księżówki, where he had the honour of meeting Karol Wojtyła, a young priest who had just come back from his studies at Vatican. It turned out they had common friends and acquaintances from the Rapsodic Theatre. The future Holy Father displayed a keen interest in Witold Kieżun's experience of the Soviet Gulag. Their paths were to cross again in 1990 in Burundi. The Pope remembered this meeting in the years gone by and their trek to Kasprowy Wierch perfectly well (Kieżun, 2013, p. 281).

And finally, I shall reveal the third reason for the Professor's cordial relations with his Alma Mater. As he recalls, it was the moment when – "having returned from the gulag and explained the Jagiellonian University Chancery the reasons of my two-years long absence, I became an object of widespread kindness" (Kieżun, 2013, p. 279).

After graduating from the Jagiellonian University, he started working in the National Bank of Poland and commenced doctoral studies under Professor Zieleniewski. He wrote a doctoral dissertation on rationalizing manager's work, obtaining a PhD degree at the Main School of Planning and Statistics (SGPiS) in 1964. Five years later, he successfully completed the habilitation procedure and obtained the title of professor. When Professor Zieleniewski retired in 1971, Professor Kieżun was appointed as a head of the Department of Praxeology of the Polish Academy of Sciences. Two years later, to the motion of the local cell (POP) of the Polish United Workers' Party, he was removed from this post to commence his work at the University of Warsaw as the Chair of the Department of Organization Theory.

In 1975, he obtained the scientific title of Professor. Since 1980, he lectured at American and Canadian universities (mainly in Philadelphia, Pittsburgh, and Montreal), but also in Western Europe and in Africa. During this period, his knowledge won him an increasing international renown. He acted as a manager of an UN aid project in Rwanda and Burundi, as well as the government expert in Burkina Faso. His theoretical knowledge and practical experience greatly contributed to creating modern organizational structures.

For his achievements, Professor Kieżun has been awarded the highest distinctions, among them the Commander's Cross and Knight's Cross of the Order of Polonia Restituta, the Home Army Cross, the Warsaw Uprising Cross, and those that are awarded to but a few: the Cross of the Order Virtuti Militari and the Cross of Valour.

As a matter of fact, Professor Kieżun's biography is a ready script for a fascinating film about Polish history of the past century. It is a wonder nobody has come up with the idea of presenting the life and work of this illustrious Scholar to the wider public.

Why management?

Life choices made by Professor Kieżun have always been clear. In principle, they require no special substantiation. Considering, however, the importance of this special moment, his choice of scholarly pursuits requires a moment of reflection.

It is fitting to start with the statement that management is a young branch of science. Perhaps for this reason there are still those who do not consider it as a science. Let us remember, however, that a Nobel Prize has been awarded for achievements in this field. Admittedly, only once, but still. It was awarded to Herbert Simon for his seminal research on decision-making processes in organizations and theoretical foundations of decision-making. Professor Kieżun also referred to his work. Moreover, he participated in Herbert Simon's seminars dedicated to perfecting the work of the manager.

Analysing Professor Kieżun's character and scholarly output, one should also ask about the motives which directed him in choosing management as a field of scientific studies. He could have become a lawyer, he could have stayed in finances, yet he dedicated his entire talent and scholarly effort to management. The reasons for this state of affairs must be sought, as it seems, in the qualities of his mind and character. In his speech from 2006, entitled *On Efficient Action and Perfection*, referring to the two notions featuring in the title, he looked for common points between the spheres of sacrum and profanum. And he did find them as much in praxeology as in management. He started by confronting the Taylorian organization, treating "the human as a machine devoid of individual initiative (Kieżun 2006, p. 16) with human relations theories", according to which people are "granted free will in the choice of the nature of their activities" (Kieżun 2006, p. 16).

In this speech, the Professor quoted Herbert Simon, who defined management as a process of permanent decision-making which cannot be optimized. According to Professor Kieżun, a perfect decision is merely a decision that satisfies us, although it is not optimal, since we are unable to reach a perfect one. In terms of the problem of efficiency of action, he invokes another of his masters, Professor Tadeusz Kotarbiński, and a close collaborator and friend, Professor Wojciech Gasparski, who formulated the "3Es" principle: effectiveness, economy, and ethicality – thus emphasizing that it is impossible to speak of effective action without vesting action with ethicality. Professor Gasparski points to the contemporary management concepts, among which he mentions management by hope, since they refer to the vision of the future, but he also speaks of the stimulation of dynamics, media influences, reinforcement of that which Professor Kotarbiński labelled as vigour – one of the fundamental characteristics of a trustworthy guardian. Professor Kieżun's speech contains a thought that can be interpreted as his peculiar credo: "…historical practice confirms that the passion for achieving goals of great format, stimulating the sphere of emotional and social involvement (…) with the visions of the future, constitutes the basis of the efficiency success in an individual and collective action" (Kieżun 2006, p. 16).

I was supposed to speak about the Polish praxeologists' contribution to the science of organization and management. It is a difficult task, since it should be the other way round, as according to Tadeusz Kotarbiński, praxeology is not part of management, but vice versa. According to T. Kotarbiński (1975, p. 480), praxeology is "evaluations common for all occupations. From that we derive: evaluations common for all types of action (a), which lead us to evaluations characteristic for various types of action (b)" (Kotarbiński 1975, p. 473).

Recapitulation

Throughout the 650 years of its existence, our University has always imparted sagacious teachings, but also propagated knowledge of importance to both Poland and Europe. Nowadays, when relations between nations are characterized by wickedness, while the peace between them seems uncertain, it is worth invoking the universal science, including the science and knowledge flowing from the Jagiellonian University, to put our own house in order.

Thus, six centuries ago, the Rector of our most illustrious Academy, Stanisław of Skarbimierz, in the forty sixth of his *Sermones sapientiales*, proclaimed that "wisdom is better than weapons of war." He also indicated "…only such a commonwealth is well-governed which in its government uses statutes rooted in the Devine Law, for the state is a God's state when its driving force is truth, when love of one's neighbours is its law, and when justice and equality are its rules" (Stanisław of Skarbimierz, 2006).

This message from our Rector, Stanisław of Skarbimierz, can be fine-tuned with Professor Kieżun's words delivered in 2006: "It turns out that our historical practice confirms that the passion for achieving goals of great format, stimulating the sphere of emotional and social involvement with strategic visions of the future constitutes the basis of the efficiency success in an individual and collective action. It turns out that we possess a rich methodological apparatus which could be enriched still further if we were able to ensure a peculiar ecumenism of the spheres of sacrum and profanum" (*On Efficient Action and Perfection*; Professor W. Kieżun's speech delivered at his Doctor Honoris Causa Award Ceremony at the Leon Koźmiński School of Business and Management on 23 May 2006).

It is worth remembering both the truths from past centuries and those from recent years, as they should be of importance for those who recognize faith as an important premise of our life and for those who think otherwise. I quoted the words of our Rector from six hundred years ago, for I know that similar ideas guided the actions of our graduate, who is now our Doctor Honoris Causa, Professor Witold Kieżun. Thanks to these words, the Professor's momentous scholarly work, his unbound patriotism, his wisdom, but also noble-mindedness, great qualities of heart and mind will forever be inscribed in the annals of the Jagiellonian University. For us, they will serve as signposts, for as an academic community we owe our graduate, Professor Witold Kieżun, gratitude for his attitude and actions bringing yet greater glory to his Alma Mater. Let the highest university honour – the honourary doctorate bestowed on Professor Witold Kieżun during the ceremonial session of the Senate of the Jagiellonian University on the 650th anniversary of its foundation – come as an expression of this gratitude.

References

Kieżun W. (2013), *Magdulka i cały świat*, Wydawnictwo Iskry, Warszawa.

Kieżun W (2012), *Patologia transformacji*, Wydawnictwo Poltext Warszawa.

Kieżun W. (2011), *Drogi i bezdroża polskich przemian*, Ekotv sp. z o.o., Warszawa.

Kieżun W. (2008), *Niezapomniane twarze*, Wydawnictwo EKOTV, Warszawa.

Kieżun W. (2006), *O dobrej robocie i doskonałości*, (In:) MBA, No. 3(80), p. 16.

Kotarbiński T. (1969), *Zagadnienie produkcyjności usług*, Warszawa, No. 33.

Kotarbiński T. (1975), *Traktat o dobrej robocie*, Ossolineum, Wrocław.

Kotarbiński T. (1987), *Opiekun spolegliwy; Pisma etyczne*, Ossolineum, Wrocław.

Montaigne M. (1952), *The Essays of Michel Eyquem de Montaigne*, Books of the Western World, Encyclopaedia Britannica.

Stanisław ze Skarbimierza (2006), kazanie 46. *O tym, że należy wyżej cenić mądrość niż oręż wojenny* (In:) *Wielkie mowy historii. Od Mojżesza do Napoleona*, Warszawa.

Professor Alojzy Czech, PhD
University of Economics, Katowice

Jan Zieleniewski (1901–1973) – Cracow period, early works

Abstract: The essay discusses early works of Jan Zieleniewski – the best known representative of the praxeological theory of organization – focusing especially on those written in the 1920s, when he studied philosophy. When working for a big industrial company, he gained practical knowledge of the functioning of a corporation. Many years later these juveniles of Zieleniewski were referred to by professor Witold Kieżun.

Keywords: praxeology, Jan Zieleniewski, biography, early works, management theory

Professor Jan Zieleniewski was not only a superior and scientific work supervisor for Professor Witold Kieżun, but also his moral authority (Kieżun, 2013, p. 385), which confirms that their acquaintance was not of merely formal nature. Their relationship, though unintentionally, was somehow richer, especially given their past. The Cracow-based history of their lives is a very special justification for mentioning the co-originator of praxeology in the context of Professor Kieżun's jubilee. They both studied at, and graduated from, the Jagiellonian University, they both refused – in spite of repeated proposals – to become Jagiellonian University research fellows and they both, each of them in his own way, experienced their worst life moments in Cracow. Professor Kieżun suffered treacherous arrest and transportation to the Soviet labour camp in Middle Asia in 1945 (Kieżun, 2013, pp. 163–165), while professor Zieleniewski was exposed to vile, ideologically driven aspersions against his family and especially his father as the owner and manager of the Zieleniewski factory (*Zieleniewszczacy*, pp. 15–21). Yet, it is important that today it is possible – without any pressure and prejudice – to come back to these matters, trying at the same time to take a position towards the hypothesis that the positive influence Jan Zieleniewski exerted on his associates and, in particular, on Witold Kieżun, during the time of creating the foundations of praxeology, was connected with values, high attitudes and skills gained at home, during studies and the times of gaining education, a first job and the beginnings of his public and social activities, stemming from the Cracow period of his biography. It is high time, then, to examine these periods of his life that had shaped Jan Zieleniewski as a co-originator of the praxeological theory of organization in Poland and an outstanding scholar.

The Zieleniewski family

Jan Michał Zieleniewski was born on 8 February 1901 in a well-known family of Cracow industrialists. His father Edmund (1855–1919) and his brother Leon (1842–1919) were the owners, and since 1906 shareholders and directors of the L. Zieleniewski Factory of Machines and Boilers S.A. in Cracow. In particular, Edmund became a director general of the factory, being at the same time a famous person of merit in Galicia. After the completion of his studies in technology in Chemnitz and Vienna, he returned to the blacksmith's shop founded in 1804 by his grandfather Antoni. Ludwik Zieleniewski, the creator of the company's meaning and a person who transformed the shop into a thriving manufacture, died in 1885. The family enterprise was taken over on a co-ownership basis by Leon and Edmund, who were not content with what they inherited. They moved the manufacture to Krowoderska St. where the optimal conditions for the production of machines – steam boilers for various industries, pumps, compressors and transmission facilities – were created. Further, since the moment of the gradual starting up of the production at the new facilities at Grzegórzki district, bridge structures, masts, steel headframes, lifts, cranes, turntables, railway points, dredgers and even passenger steamers were produced there. The range of manufacture was very wide indeed. Edmund took the technological side on his shoulders, while Leon was responsible for sales. In 1906, the company was transformed from a partnership to a joint-stock company. Austrian Credit Anstalt für Handel und Gewerbe, Lower Austria Escompt Association and dynamically developing Škoda Machinery Manufacturing joint-stock company became its minority shareholders. It was stressed that the "thriving development of the company was mainly Edmund's merit" (Saryusz-Zaleski, 1930, p. 270). In 1913, the company merged with the Sanok-based First Joint-Stock Society for Building Railway Cars and Machines of Galicia and then took over the falling Andrzej Prince Lubomirski Machinery Factory located at Zamarstinov in Lviv. The mergers were effected by a banking consortium with the Credit Anstalt as a leading party. The general management of merged companies remained in Edmund's hands. However, his entrepreneurial spirit was not content with management. Edmund engaged himself in social and public activities. He was a member of the National Industry Committee and Austrian Industrial Council, and in 1907 and then in 1911 a Member of the Parliament in Vienna. These positions were not sinecures for him. He was fully engaged in the struggle against centralist tendencies of Vienna in the field of economy. At the same time, however, he was aiming at increasing the level of industrialization of Galicia by influencing the policy of the Industry Committee. It may be assumed that he did not have too much time for his family and sons – older Edmund (who was a director of the Babcock-Zieleniewski Boiler Factory in Sosnowiec before WWII and who later became an editor of the Scientific and Technology Publishing Company) and younger Jan. This role was rather taken by his mother Jadwiga née Ciechanowska. The atmosphere of the home must have influenced the shaping of both young Zieleniewskis' personalities.

Education interrupted by war

Jan Zieleniewski studied in Cracow during the stormy years 1911–1923 (Archives of the University of Gdańsk). He attended the 3rd Jan Sobieski High School. The curriculum included a one-year stay in Vienna, where he completed the 4th Grade. At that time, the acts of war appeared cruel for the Zieleniewskis' Company. The Cracow-based factory was immobilized from the very beginning, the machinery was requisitioned, some of the machines were taken to Western Bohemia. There were no war orders. The workers of the company saw the spectre of hunger. Edmund Zieleniewski managed to gain the first order for grenade aperture as late as in 1915 and only then the machinery returned to the factory. But his other factories were in the front zone. The Russians, retreating in July 1915, virtually smashed the factories down, immobilizing them as well. The reconstruction was difficult and the future, in the post-war times, remained very uncertain, even more after the Zieleniewski brothers, who were running this major enterprise, died in 1919. Especially the death of Edmund, the entrepreneurial spirit of the company, a competent manager and an activist engaged in social welfare matters (Czech, 2009, pp. 98–99), constituted a big and painful loss.

Jan Zieleniewski obtained his maturity certificate with honours in autumn – on 26 November – of the memorable year 1918. Practically on the next day, he joined the 2nd Lancers Regiment as a volunteer, where he served from November 1918 to August 1919. Then he was sent out from his volunteer military service and, seeing the necessity to undertake certain employment in the nearest future, he enlisted in the One Year Course for FinalYear Students at the Academy of Commerce, a highly regarded Cracow school preparing final-year students to perform duties in various administration and business institutions. He completed the course on 26 June 1920 and obtained a certificate recommending his skills in administration and business matters. However, having a regular job was not to become his nearest future. Again, on the next day he returned to the 2nd Lancer Regiment, this time to fight against the Bolshevik onslaught. Demobilized in November 1920, he returned to Cracow and entered the university to continue his education at a higher level. These studies were also interrupted, in May 1921, with the service at the North Group staff of the military forces of the 3rd Silesian

Insurgence (Archives of the Jagiellonian University). Jan Zieleniewski took strict philosophy instead of certain practical courses and he specialized with success in the history of modern philosophy. Under the supervision of professor Witold Rubczyński, he wrote a dissertation entitled *Fiction in H. Vaihinger and D. Hume*. The dissertation was highly appraised. Jan Zieleniewski passed the final exams with honours and defended the dissertation on 27 June 1923, obtaining the doctor's degree in philosophy.

On the philosophy of fiction

Jan Zieleniewski's doctoral dissertation was so inventive that its first part, related to the fiction philosophy of Hans Vaihinger, was published in *Kwartalnik Filozoficzny* (*Philosophical Quarterly*) (1924, ch. 2). This does not mean, of course, that the second part, taking the same topic in the philosophy of David Hume, was of less importance. It happened so, because Dr. Jan Zieleniewski did not join the academic centre of philosophy and probably he did not finalize his publishing undertaking. Apart from that, the fictionalism of Vaihinger was more attracting, as it was inscribed within modernity. It was one of the most original positions in German neo-Kantism or even in European philosophy in general. It appeared as a comet, and died in an instant, like a comet, too.

A departure from metaphysics was a feature of philosophy of the second part of the 19th century. No one was interested in searching for the nature of things; a picture of the phenomenon was deemed to be sufficient. Almost all thinkers of the time agreed that there was no way to search for the absolute truth. "The world is for us, there is no world in the inside of us" – Petzold said; "truth is usefulness" – as pragmatists wanted; "life is the value of values" – Nietzsche emphasized; "the will to the foreground" – Wundt demanded; "pragmatic humanism is needed" – Schiller expostulated; "and human thinking creates things" – as neo-Kantists believed (Zieleniewski 1924, p. 159). This was the atmosphere of the era.

Vaihinger encouraged his readers to follow Friedrich A. Lange's statement that fictions are "necessary factors of human thinking, particularly, scientific thinking" (Zieleniewski 1924, p. 159). As a result, the structure of science was believed to be founded on notions that did not exist in reality. This applied to all spheres, including natural science and economy. In mathematics, it has been clear from the very beginning. We have infinitely small values, a circle as a polygon having an infinite number of sides, or a perfect line, or non-dimensional points. But we have also negative values or imaginary numbers. In fact, formal logic in general was a fiction for Vaihinger (Zieleniewski 1923, p. 16). Natural science posits an isolated object, constantly accelerated motion, atom or ether. In the methodology of economics, we can see *homo oeconomicus* and in law – *fictio iuris*. Is it possible that sciences are based on a certain fiction?

However, this observation of the condition of science did not lead Vaihinger and then Zieleniewski to skeptical positions. When tracing these fictions, they used to say that fictions were useful and practical. The question then – what are they? – provides us with the answer in the form of often lengthy philosophical works, including Jan Zieleniewski's dissertation. However, this is not a practical question. It is better to ask: how is it possible that, with the help of imaginings of which we know that they are false, we arrive at correct results? (Zieleniewski 1924, p. 160). According to Zieleniewski, fiction is a mental phenomenon that is at variance with reality. This, however, does not apply to acts of will and emotions which are excluded from the realm of fiction. Vaihinger assigned static features to fictions – as a result of logical functions. Zieleniewski, in turn, viewed this as a limitation, since fictions, he argued, are also "something that is going on" (Zieleniewski 1924, p. 173), such as e.g. operations, analogies, divisions, abstract generalizations, unallowed transformations, etc. They have a clearly dynamic characteristic. It may be assumed that the fictions are (Zieleniewski 1924, pp. 173–177):

- in contradiction or deliberately wrong (in a double way: in relation to reality or inherently contradictory, such as e.g. infinitely small value, infinite quantity);
- temporary (historical, subject to becoming obsolete);
- purposeful (they are always means leading to a goal);
- useful (when applied in thinking, they lead to the right results).

Thus, Zieleniewski makes a sudden change in the narration his dissertation in order to state that thinking is subjected to acting and enables action. Thinking is a transitional stage, it is a means to achieve a goal which is enabling practice understood as an ordinary action or ethical way of acting. Thinking is only a means to reach the goal; Kant's primacy of practical reason remained a permanent basis of this well-known standpoint.

Why, then, does the inclination of thinking to draw one away from the problems of life and activity towards taking care of himself only appears so often? It is so because, as Vaihinger answers, we have to act according the law of "means growing above goals" (Zieleniewski, 1924, p. 207) which is present everywhere, both in nature and the psychical world. The action "which was previously a means to reach the goal, becomes now a goal in itself and draws its strength from the primary goal" (Zieleniewski, 1924, p. 207; 1923, p. 76). It often happens to thinking. It was transformed from the means used for vital goals into a goal in itself, which is useless and speculative. By the way, the right of means growing above the goals was, one would say, the only conclusion from the doctor's dissertation applied to the fundamental works on organization and management (Zieleniewski, 1965, p. 215, ref. 84), where it took the form of the autonomization of auxiliary organizational units.

Placement in the industry, organizational training, first publications

Jan Zieleniewski started his job with the Cracow-based office of the "L. Zieleniewski" Factory of Machines and Railway Cars joint-stock company on 1 December 1924 in the capacity of a trainee, to become, after certain period of time, a secretary to the management board of the Cracow company (Archives of the University of Gdańsk). The firm underwent deep restructuring under the Polish regulations (see more: Czech, 2006, pp. 137–141 and 145). It merged with the Fitzner and Gamper Boiler Factory in Sosnowiec and then entered capital relations with the Friedenshütte A.G. concern, to mention only the most important changes. These changes became the inspiration and justification of one of the earliest Zieleniewski's works dealing with the concentration of production. I shall elaborate on this point below.

The work in the management board office of the concern must have revealed certain gaps in Zieleniewski's professional knowledge, even if his general education was advanced, as he took a Private Training Course in Scientific Organization of Labour organized by Curt Piorkowsky in Berlin in the summer of 1926. Simultaneously, from 1924 to 1930, Jan Zieleniewski performed the function of a Secretary of the Economic Society in Cracow, thereby familiarizing with the circle of economists – members of the Society.

Due to the direct inspiration by his observation of activities of the L. Zieleniewski and Fitzner-Gamper S.A. United Factories of Machines, Boilers and Railway Cars, deepened by theoretical studies on company mergers, the essay on concentration was one of the first publications of this type in Polish literature. The essay described the rule of horizontal and vertical mergers as well as the forms of industrial unions: cartel, syndicate, concern, trust. The description was illustrated with examples, most often coming from Upper Silesia, but also from the world industry. The author made a review of opportunities and threats of the pending process, showing its economic consequences, in particular the impact of the concentration on prices, costs and rationalized production. He appraised this influence positively which could mean that he stood for the uniting of enterprises. In this regard, he highlighted the potential to eliminate, or at least limit, competition, joint finances of united factories having different profiles of production, and beginning economic activity according to the joint plans. The property itself was of a lesser importance. In the end of the work, the author remarked on the evolution of organizational forms, which his mother company had been subject to, showing that this was an example of the vertical-horizontal concentration characteristic of the metal processing industry (Zieleniewski, 1929, p. 60).

In 1929, the Presidium of the Government appointed a Committee for Streamlining Public Administration, acting in six sections, led by professor Maurycy Z. Jaroszyński (1890–1974). Jan Zieleniewski was a member of the Committee. From 1929 to 1930, he worked as a secretary in the section of cash procedures and materials management.

Antoni Lewalski (1878–1941), a president of the management board since the death of the Zieleniewski brothers in 1919, was a head of this section; he was a mentor of the young employee and wanted to have "his man" there. Jan Zieleniewski's duties included making reports on government audits, budgeting, public accounting and cash operations, materials management and wages (Archives of the University of Gdańsk). As it seems, accounting became particularly close to Zieleniewski. He wrote *Guidelines for Public Accounting*, a very special document, in which the Author's fascination with the idea of accounting, and with the rules governing accounting, was clearly expressed. Inspired by great figures in the history of culture and science, he drew a conclusion from the analogy between accounting transactions and mechanical movement, claiming that this field should be treated as a science. For Jan Zieleniewski, accounting is a science based on fictions. "Economic values are variable: work of a labourer or a merchant or a change of market situation give goods greater real value than the sum of raw materials and labour used for processing, transport, etc. Nevertheless, accounting in a way uses a fiction of indestructibility of economic values. When, as a result of a change of price situation, a surplus appears on the account of certain stock of goods, i.e. a new value which in reality came from nowhere and which was not given to us by someone, the commercial accounting shows this phenomenon as if (like Vaihinger's fiction) this new value was given to the enterprise by its owner. When consuming (thwarting) of certain values takes place in such a way that new values do not appear in their place (for example, spending Budget funds for pensions), accounting creates a fiction: it shows the consumed value as the amount that budgetary accounting increases, as the account of pensions. It debits this account, crediting at the same time the account on which this amount could be found previously (cash account)" (Zieleniewski, 1930, p. 20).

Let us take another example. Why are profits – which are increasing the capital – booked on the right side which is used for showing losses in all kinds of commercial accounting? It is simple: they are evidenced on the appropriate accounts of property elements on the left side. Evidencing them on the capital accounts expresses the fiction of invariability of the sum of values, showing profits as the value with which the enterprise is entrusted by its owner. Zieleniewski stresses that this fiction is particularly fertile in the accounting. It makes the systematic accounting have internal mechanism of automatic control. If we present the values created within the business unit as the values obtained from the outside (fiction again), and the thwarted values as the values given to someone standing outside this business unit – then this is the mechanism that makes it possible to consistently apply the primary rule of double booking system, i.e. evidence each transaction first on a credit and second on a debit side of another account. At the same time, this is the guarantee that the sum of debit balances of all accounts must be always equal to the credit balances; all accounts are in the state of balance (Zieleniewski, 1930, pp. 20–21). This a bit lengthy reasoning clearly illustrates how much can be seen in the field into which one comes from the outside. What is more – after the course in organization in Berlin – Zieleniewski learned accounting as part of the whole science on the enterprise, referring to the forerunners of *Betriebswirtschaftslehre*: Leo Gomberg (1866–1935) and Johann F. Schär (1846–1924), both of them being noticeable with their academic achievements in the field of accounting.

As far as the general part of the *Guidelines* is concerned, the paper by Zieleniewki described several groups. The second part was devoted to organizational rules. The duty of regularity was reminded and then main problems of public accounting in Poland were presented: with respect to property status, standards of economy, receivables and obligations. The author discussed the organization of accounting divided into branch and territorial accounting. In this methodology of accounting, a basic problem appeared: the chart of accounts. In the fields of techniques: the today forgotten issues of lines in the accounting sheets. The end of the paper contained a draft of some *ad hoc* reforms related to budgetary income and expenditures in the Treasury and outside of the Treasury branch. This very active participation in the works of the government committee was instructive for the author, but also highly valued. Jan Zieleniewski was granted the golden cross of merit in appreciation of his contribution to the works streamlining functioning of the state.

Summary

After the activities of the Warsaw Committee had been closed, Jan Zieleniewski did not continue his professional career in Cracow. In virtue of a decision made by the new management board, he obtained a position of the head of a small company – the W. Fitzner Factory of Boilers and Welded Products Ltd. in Siemianowice.

The factory was owned by Zieleniewski & the Fitzner-Gamper concern that had, among other things, innovated welding technologies. Zieleniewski ran this company from 1930 to 1934, trying to keep involved in his extra activities from the Cracow period. Very soon he identified himself with aims and interests of Silesian industry, turning to the reconstruction of structures and the directions of development of the large-scale industry. Zieleniewski was involved in the Maritime and Colonial League that was very active in Silesia as a society promoting the opening of Poland to the world through the sea. The Silesian engineers and economists' circles were firm proponents of the hard coal pro-export policy; this raw material took over the position of cereals which had been the main export good in old Poland. In his review, Zieleniewski underlined the accessibility and solid documentation of the argument. The last publication of his was a dissertation praising the solutions applied in the insurance business that he had seen in Silesia. The early works of Jan Zieleniewski were exhausted at this point. This astonishing fact can be explained by his growing duties in Upper Silesian industry. From 1934 until the outbreak of WWII, he was involved in the largest undertakings as a secretary to the management board of the Wspólnota Interesów S.A. – the Katowice-based Company for Mining and Metallurgy and United Steelorks, as well as an enforced administrator of Goods and Industrial Works of the Prince of Pszczyna and the Hohenlohe Works S.A. However, this fascinating and unknown experience of Jan Zieleniewski as a manager of huge companies may be a topic for another paper.

References

Archives of the University of Gdańsk, file Z – 13.

Archives of the Jagiellonian University, file WF II 504.

Czech, A. (2006), Fuzje i przejęcia? Ależ to było zawsze – przykład z dziejów przemysłu polskiego, *Zeszyty Naukowe AE w Katowicach*, No. 37.

Czech, A. (2009), Edmund Zieleniewski – przedsiębiorca prospołeczny, *Współczesne zarządzanie*, No. 1.

Ekonomista 1933, Vol. I.

Hume, D. (1974) *Mój żywot*, (In:) S. Jedynak: *Hume*, WP, Warszawa.

Kieżun, W. (2013), *Magdulka i cały świat*, Iskry, Warszawa.

Przegląd Współczesny (1929), No. 88, August.

Saryusz-Zaleski W. (1930), *Dzieje przemysłu w b. Galicji 1804–1929 ze szczególnym uwzględnieniem historii rozwoju S.A. L. Zieleniewski i Fitzner-Gamper*, Cracow.

The speech of the Head of Praxeology Chair, Associate professor Witold Kieżun (1972), *Prakseologia*, No. 44.

Zieleniewski, J. (1923), *Fikcja u H. Vaihingera i D. Hume'a* (hard copy of the doctoral dissertation No. 878), UJ, Kraków.

Zieleniewski, J.(1924), *O "Filozofii fikcji" H. Vaihingera*, "Kwartalnik Filozoficzny", book II.

Zieleniewski, J. (1929), *Koncentracja produkcji*, by the Economic Society in Cracow, Warsaw.

Zieleniewski, J. (1930), *Wytyczne rachunkowości publicznej*, by the Industrial and Commercial Chamber, Warsaw.

Zieleniewski, J. (1965), *Organizacja zespołów ludzkich. Wstęp do teorii organizacji i kierowania*, 2nd Edition, PWN, Warszawa.

Zieleniewszczacy. 25-lecie strajku 1936–1961 (1961), Kraków.

Part II
Reflective Management Dilemmas

Part II

Professor Łukasz Sułkowski, PhD
Uniwersytet Jagielloński
Społeczna Akademia Nauk

Metaparadigmatic perspective in the theory of organizations and management sciences

Abstract: The chapter presents the meta-paradigmatic perspective on the theory of organizations and management science. At the beginning, it defines paradigms in social sciences and management and reviews different methods of their classification. Then, a meta-paradigmatic approach, which would overcome the weaknesses of the approaches based on a single cognitive perspective, is proposed.
Keywords: paradigms, social sciences, management, cognitive perspectives, meta-paradigmatic reflection

Introduction

Different ways of understanding paradigms in social sciences, and particularly in the discourse of organization and management, lead to seeking a cognitive position which will embrace the perspective of a single paradigm (Gioia & Pitre, 1990, pp. 548–602). Such a concept is a metaparadigmatic approach which allows looking at the same issues of organization and management from the perspective of different basic assumptions, theories and research methods (Sułkowski, 2012). The aim of the article is to present this kind of meta-paradigmatic perspective on the theory of organization and management science. In order to unfold this concept, it is necessary to define paradigms in social sciences and management and to review the methods with which they are classified. Then, a meta-paradigmatic approach, which would overcome the weaknesses of approaches based on a single cognitive perspective, will be proposed. Adopting the metaparadigmatic perspective strengthens the epistemological and methodological pluralism in the discourse of organization and management. Nonetheless, it needs to deal with the problem of incommensurability of theories and methods stemming from different research traditions (Poropat, 2000, p. 325).

Meaning of a paradigm in management sciences

The definition of paradigm in social sciences is far from being clear. Thomas Kuhn himself – a classic who famously popularized the idea of paradigms in science –

employed many different meanings of the word "paradigm" in science. The most common definition of paradigm is a "universally recognized scientific achievement that, for a time, provide model problems and solutions for a community of practitioners" (Kuhn, 1996, p. 10).

Among the criteria of scientificity defining every paradigm, a few epistemological aspects can be pointed out (Kuhn, 1996, p. 10):

- What is to be observed and studied?
- What kind of questions should be asked and answers given in relation to the subject of the research?
- How research results should be interpreted?
- How should research be carried out?

From the perspective of management sciences, it seems rational to assume that a paradigm covers the basic epistemological and methodological assumptions of a given scientific discipline, assumptions, which are accepted by a community of researchers. The paradigm is thus a historically varying *consensus omnium* of specialists belonging to a specific scientific discipline. Thomas Kuhn noticed certain historical variability in scientific research and issues in different areas and disciplines. From a historical perspective, science develops primarily through the accumulation of achievements, that is, the development of research and theory, but the most important scientific advancements are also scientific revolutions during which a revision of the foundations of science is carried out. L. Laudan describes the complexity of the process of scientific progress, which is far from a simple accumulation of knowledge postulated by logical empiricism (Laudan, 1977). T. Kuhn called cumulative periods of progress normal science, describing the time of changes of cognitive foundations as the scientific revolution. In social sciences, it is difficult to speak of a dominant paradigm, as they tend to rely upon different, disproportionate, and sometimes contradictory paradigms. And so, in the theory of organizations many possible concepts of division of paradigms and a parallel operation of different scientific schools and paradigms can be identified, which suggests that we are dealing with a multiparadigmatic science.

Kuhn's model, cited here, and later altered by I. Lakatos, can turn out to be useful for looking at the subsequent stages of development of the theory of organization and management sciences. Then, the question arises as to on which stage management sciences are today. This question cannot be given a definite answer. Some researchers locate management sciences at a pre-paradigmatic stage, which translates into the fact that the dominant cognitive perspective cannot be drawn yet. Still, others insist on the permanent coexistence of many paradigms, meaning the functioning of numerous methodological and epistemological assumptions (Sułkowski, 2012).

The theory of organizations and management sciences uses multiple paradigms; there are also different ways of their differentiation. The vast majority of researchers assume a single cognitive perspective, adopting an approach to organization and management which can be described as classical or alternative (i.e. postmodernist and critical). If we consider the epistemological awareness, we can say that researchers adopt a paradigm reflexively (explicitly) or implicitly. An increased reflexivity often leads to reaching for alternative paradigms. The multiparadigmatic or metaparadigmatic approach or a complete abandonment of a paradigm in organization and management theory are rarer (Miller, 2007, pp. 177–184).

Divisions into paradigms in the theory of organizations and management sciences

Management theory jungle results from the complexity of the studied issues, which is characteristic of social sciences and reflected in the multiplicity of paradigms. Various researchers exploit the incommensurable concepts of organization and management paradigms, out of which 8 approaches have been selected. They can be ordered according to their prevalence in the scientific community (According to Google Citation, December 2014):

1. Gareth Morgan's organizational metaphors (more than 15.000 citations) (Morgan, Gregory & Roach, 1997).
2. Organizational research paradigms by Gibson Burrell and Gareth Morgan (more than 10.000 citations) (Burrell & Morgan, 1979).
3. Cognitive frames of understanding organizations by L.G. Bolman and T.E. Deal (more than 6.500 citations) (Bolman & Deal, 1991).
4. Management paradigms presented by Mary Jo Hatch (more than 4.000 citations) (Hatch, 2012).
5. System paradigms of management by M.C. Jackson (more than 3.000 citations) (Jackson, 2000).
6. Epistemologies of management research by P. Johnson and J. Duberly (more than 800 citations) (Johnson, Duberley, 2000).
7. Classical approach of management schools in accordance with chronology.
8. Classical approach of organization and management issues in accordance with the division into management areas (or subdisciplines).

Table 1. Classifications of paradigms in the theory of organizations and management sciences.

Classification of paradigms	Author/Authors	Ordering rule (meta-rule)	Type of preferred epistemology	Type of preferred methodology
Organizational metaphors	G. Morgan	Analogies and multiple metaphors	Interpretative, hermeneutic, pluralism	Qualitative, seeking for sense
Organizational research paradigms	G. Burrell, G. Morgan	Approach to organizational reality and to organizational changes	Epistemological pluralism – identification of 4 models of perfect paradigms	Methodological pluralism, emphasis on qualitative methods
Cognitive frames of organizations	L.G. Bolman T. E. Deal	Analogies and multiple comparisons	Epistemological pluralism – differentiation of cognitive perspectives	Equivalence of methodologies
Systems of organization and management	M.C. Jackson	Universality of open system approach	Epistemological pluralism – multiplicity of systems	Methodological pluralism
Epistemologies of research of organization and management	P. Johnson and J. Duberly	Ontology and epistemology of organizations as fundamental dimensions	Epistemological pluralism – identification of 5 models of perfect paradigms	Methodological pluralism

Classification of paradigms	Author/ Authors	Ordering rule (meta-rule)	Type of preferred epistemology	Type of preferred methodology
Schools of management	H. Mintzberg, M. Bielski, Wren, Daniel A. and Arthur G. Bedeian	Chronology, scientific centres and authorities developing schools of thought	Epistemological fundamentalism, cumulative concept of knowledge development	Preference for quantitative methods in relation with their rigorism
Division into areas of management (or subdisciplines).	W.M. Grudzewski, S. Sudoł	Division into key areas of organization and management which can develop scientific subdisciplines	Epistemological fundamentalism, cumulative concept of knowledge development	Preference for quantitative methods in relation with their rigourism

Source: author's own study.

G. Morgan's organizational metaphors

G. Morgan presents another way of understanding the epistemology of management. This conception is different than the juxtaposition of paradigms suggested by himself and G. Burrell. The researcher proves that understanding of management is made possible thanks to a specific metaphor of an organization that determines the actions of both a member of the organization and the researcher (Morgan, 2007). The pluralist understanding of organizations is facilitated by eight metaphors which highlight different characteristics. It has to be admitted that the organizing metaphor used by researchers and managers allows one to include both reflections and actions in the conceptual and pragmatic framework. In thinking through the metaphor (whether cultural, organic, mechanistic, cognitive or psychodynamic), stereotypes and various types of management discourses are manifested. However, for G. Morgan the organizational metaphor is not the same as management paradigms. The latter should be characterized by a higher level of generality. What is more, they are required to have an impact on the achievements of a given science.

G. Burrell and G. Morgan's organizational paradigms

The approach by G. Burrell and G. Morgan defining four paradigms of management (and dominating in the social sciences) belongs to the most popular ones (Burrel & Morgan, 1979). Assumptions associated with cognition (objective and subjective) and social orientation (regulation, change) can be indicated as criteria for the differentiation of paradigms. If the above features are crossed with each other, we will receive four different paradigms: functionalist, radical structuralism, interpretative, and radical humanism. The modified approach of these researchers is presented in Table 2 (see also Sułkowski, 2012).

Table 2. Paradigms existing in social sciences

Epistemological perspectives on science		Choice of social orientation	
		Regulation	Radical change
	Objectivism	Functionalism	Radical structuralism
	Subjectivism	Interpretative-symbolic paradigm	Postmodernism

Source: on the basis of Burrell & Morgan, 1979.

Cognitive frames of the organization by L.G. Bolman and T.E. Deal

Other authors, L.G. Bolman and T.E. Deal, put forward the concept of "cognitive frames" to enable a comprehensive analysis of organizational management. Structural, political, symbolic and human resources management approaches belong to this framework (Bolman & Deal, 2003). These frames combine both a prevailing look at paradigms in social sciences (the symbolic approach and structuralism) and the subdisciplinary approach (human resources management and political approaches). The authors cite several management problems as examples of the application of this scheme. Among the drawbacks of this concept, one can indicate a lack of rooting in the discourse of other social sciences. The described frames also appear to be insufficient to cover all cognitive perspectives of management. The scheme does not leave space, for example, for the neo-revolutionary approach, critical trends in management, and the incremental concepts of strategy.

Four management paradigms by M.J. Hatch

M.J. Hatch distinguishes between four paradigms of management sciences, using Burrell's and Morgan's proposals and a historical perspective. These paradigms include: classical, modernist, interpretative-symbolic, and postmodernist. Their main characteristics is presented in the table below.

Table 3. Cognitive perspectives of organization and management by M.J. Hatch

Perspective	Subject	Result	Method
Classical	Organizational management, Influence of organizations on management	Conclusions for the practice of management, Theoretical schemes	Observation and historical analysis, One's reflection on the basis of experience
Modernist	"Objective measures" are used to analyse organizations	Statistical analyses with the application of many variables, Comparative studies	Correlation of standardized methods Descriptive methods
Symbolic-interpretative	"Subjective review" is used to look at organizations	Organizational ethnography, Descriptions of cases, narration	Ethnographic interviews Participating observation
Postmodernist	Theory of organization Practices of theorizing	Reflection	Deconstruction Criticism of theoretical research

Source: Hatch 2002, p. 63.

The classical paradigm, described by M. Weber, H. Fayol, F.W. Taylor and Ch. Barnard, has its cognitive foundations in neopositivism (Sułkowski, 2004). Modernism (represented by authors such as H. Simon, L. Bertalanffy and J. March) recognizes organization in the system and functionalist categories (March & Simon,

1964). P. Berger, P. Seleznick, T. Luckman and E. Hoffman are representatives of the interpretative-symbolic approach, perceiving the organization as a process of creating and interpreting social reality of the organization (Berger, Luckmann, 2010). Postmodernism draws attention to cultural and epistemological relativism and defragmentation which appear in narrative and textual frames (Hatch, 2002, p. 204–238). Representatives of postmodernism include K. Dale, N. Monin, B. Czarniawska-Jorges, G. Burrell (Burrell & Dale, 2002). Paradigms, according to M.J. Hatch, are characteristic of management and refer to different stages in the development of discourse, but they have no place for the critical trend, which is recognized by the authors of *Critical Management Studies* as a separate paradigm.

System paradigms of management by M.C. Jackson

The system approach to organizations aspired to the role of the dominant paradigm of organization and management for three decades. System models were to be the most general categories describing the functioning of the organization. However, the system approach has been questioned since the 70s of the twentieth century and often identified with the neopositivist and functionalist approaches, which are static and do not reflect the complexity of the organization's social life. The interpretative, postmodernist and critical approaches question the value of system models. M.C. Jackson suggests the possibility of a systemic approach in its broadest sense. He claims that functionalist models of the organizational system are just a special case of the theory of organizational systems. Jackson proposed the concepts of "loose" organizational systems that can be identified with alternative paradigms of organization and management theory (Jackson, 2000).

Epistemologies of management research by P. Johnson and J. Duberley

The perception of management research epistemology by P. Johnson and J. Duberley reaches to philosophy and differentiates five types of reflection in the study of management, combining the ontological and the epistemological dimensions. The proposal of these researchers is illustrated in Figure 1.

Figure 1. Between ontology and epistemology – P. Johnson and J. Duberley

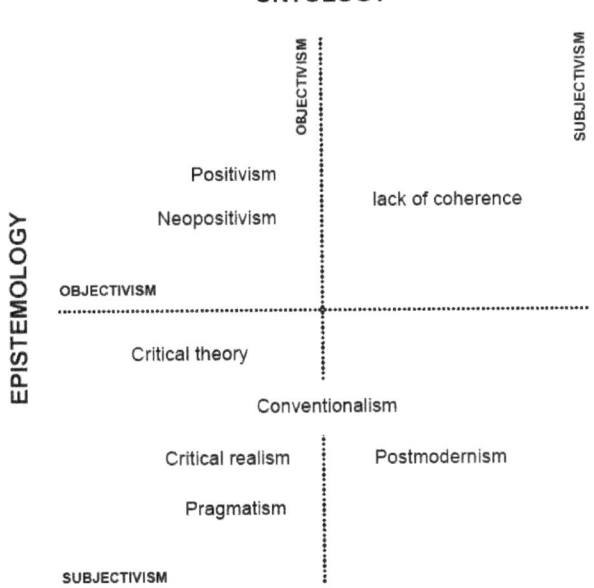

Source: Johnson & Duberley, 2005, p. 180.

The concepts presented by the above mentioned authors bring new features (compared to Burrell and Morgan's theory): positivism, neopositivism, conventionalism and pragmatism (Johnson & Duberley, 2005, pp. 180–191). Objective ontology and epistemology, characteristic of positivism and neopositivism, are equivalent to functionalism (from the typology by G. Burrell and G. Morgan), and at the same time to the classical and modernist approach (concept by M.J. Hatch). Conventionalism, in turn, situated (like postmodernism) between subjective ontology and epistemology, provides greater reflexivity due to its reaching for the untranslatable metaphors and paradigms of organization and management.

Pragmatism and the critical trend mean an involved knowing and the desire to describe the reality, that is, they operate within the subjectivist epistemology and the objectivist ontology. This concept complements the previously described paradigms by Burrell and Morgan, if we consider the reflexivity of research in management. It must be admitted, however, that this approach still bears the hallmarks of a novelty and is not widespread within management sciences, and – more broadly – in the context of social sciences.

Key areas of management

The issue of deepening specializations can be observed in management sciences. Within their domain, there are many subdisciplines (from a few to a few tens), benefiting from diversified research methods, differently comprehending the subject of study and representing different research approaches (Sudoł, 1999). These subdisciplines shape their own discourses, which do not always complement each other, they remain related in varying degrees with the sciences which are an inspiration to management. The following, not entirely separate, can be listed among these subdisciplines (naturally, it is not a complete list): strategic management, financial management, managerial accounting, human resource management, public management, marketing management, organizational theory, quality management, entrepreneurship, business science, international management, production management, leadership, innovation, information management, change management in the organization, organizational communication, management methods and techniques.

Such a division is associated with the existence of more narrow paradigms and, consequently, the work of narrower groups of experts working on the solution to a given research problem. Each management area (be it human resource management, marketing or financial management) would have its own cognitive assumptions, that is, it would become a separate paradigm. This solution does not fit in the broad definition of paradigm, which assumes certain consensus in the context of a larger group of researchers. Moreover, in the referenced subdisciplines one can also indicate different cognitive assumptions (for example, a functionalist or an interpretative paradigm of organizational culture). It is worth mentioning that the division of management subdisciplines is not constant, it is conventional and, what is more – remains controversial in the academic environment (e.g. Sudoł & Kożuch, 2010). For these reasons, it is difficult to treat the subdisciplines of management as paradigms of this science.

Management schools by chronology

There have been many trends and schools bringing new content to the discourse, but also introducing more complexity to the problem of organization and management (Koontz, 1961). Chronological perspective, which allows to look at the development of a variety of schools in the field of management, is proposed by many researchers (Mintzberg, 2004). M. Bielski indicates fourteen tendencies of the three trends (classical and neoclassical, psycho-sociological and systemic) (Bielski, 1996, p. 42). For example, the basic assumptions of the school of social relations by E. Mayo, scientific management postulated by F.W. Taylor, or administration school promoted by H. Fayol are different. Common to the three aforementioned schools are: their understanding of the truth, the realm of values, and ontology. Similarly, one could analyse other trends, classical, neoclassical, systemic, psychosocial. Thus (as has been shown in the case of subdisciplines), a more general criterion for the differentiation of paradigms seems necessary.

Metaparadigmatic reflection

The choice of the paradigm is made by means of adopting certain ontological, epistemological and axiological principles in the organizational reality by management theoreticians and researchers. Numerous management theories can be embedded in various paradigms. What lies at the heart of the metaparadigmatic reflection is a reflective juxtaposition of the results of research and interpretations taken from various paradigms or even from different typologies of paradigms. Due to the wide variety of the ways of understanding and exploring organizations, one has to consider the relationships between management "paradigms". There are a few possible interrelations between the paradigms:

1. The conflict of paradigms.
2. The incommensurability of paradigms.
3. The integration of paradigms.
4. The hybridization of paradigms.

From the logical viewpoint, it is usually impossible that two paradigms should be true, since they are based on contradictory assumptions. The concepts are antithetic; they form paradoxes and antinomies whose mutual relations can be based on criticism of the fundamental assumptions. The case of contradictory paradigms is fairly common in management studies. Taylorism and the school of social relations are rooted in different visions of human nature in the process of organizing. Modernism and postmodernism perceive management ontology, epistemology and axiology in radically different ways.

This incommensurability denotes the untranslatability of the concept. The discussion on the incommensurability of paradigms was initiated by T. Kuhn (2001). The radical standpoint postulating the impossibility of rational argumentation, reaching compromise or even communication is represented by the cognitive relativists, e.g. P. Feyerabend, the Edinburgh School or postmodernists (Feyerabend, 1979). The presented examples of the juxtapositions – Taylorism versus the school of social relations or modernism versus postmodernism – indicate radically contradictory assumptions which can yet be a subject of comparison.

Incommensurability is the case of completely different, incomparable concepts of the organizational reality.

Integration denotes combining paradigms by means of searching for common points and leaving contentious issues open to be solved (Gioia & Pitre, 1990). G. Morgan noticed that the coexistence of a few paradigms, thanks to the synergetic effect, offers new possibilities of development for social sciences and organizational studies (Morgan, 1983). Such integration is possible on the basis of conventionalism, since it is possible to assume that in the research we do not investigate the heart of the matter but we expose various points of view (what serves as a philosophical basis here is antiessentialism and subjectivism), e.g., the integration of various organizational metaphors can be the source of knowledge for both the manager and the scholar.

It is also possible to hybridize paradigms. Some cognitive attitudes can offer creative inspirations for other approaches. An example of such an interplay between paradigms is the hybridization of certain ideas of organizational culture on the basis of the functionalist and the interpretive paradigms based on the use of the critical tools of postmodernism (Schultz & Hatch 1996). By definition, paradigms should be incomparable or contradictory. Still, it seems that it is possible to juxtapose them and indicate the differences, which can lead to the increase of reflexivity, thereby providing a deeper understanding of the phenomenon of organization and management (Schultz & Hatch, 1996).

Conclusions

By definition, paradigms are incomparable or contradictory because they are based on different basic epistemological or methodological assumptions. It is seems possible, however, to compare, juxtapose or indicate differences, which can lead to the increase of reflexivity and, consequently, to a better understanding of the phenomenon of organization and management. A manifestation of this position can be seen in the metaparadigmatic approach which allows a multiplicity of attitudes from the perspective of different paradigms, and even different ways of classification of paradigms in management sciences. While seeking the foundations of the metaparadigmatic approach in the theory of organizations and management studies, one has to assume that:

1. A paradigm is a *consensus omnium* referring to cognitive foundations accepted by groups of researchers and representatives of scientific schools, and not by the whole of the scientific environment (Amsterdamski, 1992, pp. 9–26).

2. A paradigm covers common postulates referring to: the nature of organizational reality (ontology), the way of knowing it (epistemology and methodology), the criteria of scientific truth (correspondence, coherent, constructivist), the attitude to values (axiology), the identity of researchers (identity of a scientific school and scientific institutions), and the approach to management practice (pragmatism).
3. The multiparadigmatic and metaparadigmatic approach, based on epistemological and methodological pluralism, is allowed (Lewis & Kelemen, 2002, pp. 251–275).
4. Methodological triangulation in the research area is postulated (Cox & Hassard, 2005, pp. 109–133).
5. Incommensurability, and even contradiction, of the results of research of organization and management problems is acceptable (Slattery, 2003, pp. 151–155).
6. Reflexivity and deepening the epistemological awareness of the researchers are desired (Alvesson, 2003, pp. 13–33).
7. Poly-methodology, which allows one to use varied quantitative and qualitative methods, is advisable (Mingers & Gill, 1997).

References

Berger, P.L., Luckmann, T., (2010), *Społeczne tworzenie rzeczywistości*, PWN.

Bolman, L.G., and Deal, T.E. (1991), *Reframing organizations*. Vol. 130. San Francisco: Jossey-Bass.

Bolman, L.G., and Deal, T.E. (2013), *Reframing organizations: Artistry, choice, and leadership*. John Wiley & Sons.

Bolman, L.G., Deal, T.E. (2003), *Reframing Organizations. Artistry, Choice and Leadership*, Jossey-Bass, San Francisco.

Burrel, G., Morgan, G. (1979), *Sociological Paradigms and Organizational Analysis*, Heinemann, London.

Burrell, G., and Morgan, G. (1979), *Sociological paradigms and organizational analysis*. Vol. 248. London: Heinemann.

Burrell, G., Dale, K. (2002), *Utopiary: utopias, gardens and organization*, (In:) M. Parker (Ed.) Utopia and Organization, Blackwell, Oxford.

Gioia, D.A., and Pitre E. (1990), Multiparadigm perspectives on theory building. *Academy of Management Review 15(4)*.

Hatch, M.J. (2012), *Organization theory: modern, symbolic and postmodern perspectives*. Oxford University Press.

Hatch, M.J. (2002), *Teoria organizacji*, PWN, Warszawa.

Jackson, M.C. (2000), Systems approaches to management. Springer.

Jackson, M.C. (2003), *Systems thinking: Creative holism for managers*. Wiley & Sons: Chichester.

Jackson, M.C. (1991), *Systems methodology for the management sciences*. Springer.

Johnson, P., Duberley, J. (2003), Reflexivity in Management Research. *Journal of management studies*, 40(5).

Johnson, P., Duberley, J. (2005), *Understanding Management Research*, Sage, London.

Johnson, P., Duberley, J. (2000), *Understanding management research: An introduction to epistemology*. Sage.

Koontz, H. (1961), The management Theory Jungle, *Journal of the Academy of Management*, 4(3).

Kuhn, T.S. (1996), *The Structure of Scientific Revolution, The Structure of Scientific Revolutions*, 3rd edition. University of Chicago Press, Chicago.

Laudan, L., (1977), *Progress and Its Problems: Towards a Theory of Scientific Growth*. University of California Press, Berkeley.

March, J.G., Simon, H.A. (1964), *Teoria organizacji*, PWN, Warszawa.

McAuley, J., Duberley, J., Johnson, P. (2007), *Organization theory: Challenges and perspectives*. Pearson Education.

Miller, D. (2007), *Paradigm prison, or in praise of atheoretic research*, Strategic Organization, 5.

Mintzberg, H. (2004), *Managers, not MBAs: A hard look at the soft practice of managing and management development*. Berrett-Koehler Publishers.

Morgan G., Gregory F., Roach C., 1997, Images of Organization, Sage, Thousand Oaks.

Morgan, G. (1997), *Obrazy organizacji*, PWN, Warszawa.

Poropat, A.(2000), *Comparing apples with oranges: Teaching and the issue of incommensurability in organizational studies*. Proceedings of the conference Transcending Boundaries: Integrating people, processes and systems. M. Sheehan, S. Ramsay & J. Patrick (Eds.) Griffith University, Brisbane.

Sudoł, S. (1999), *Przedsiębiorstwo: podstawy nauki o przedsiębiorstwie: teorie i praktyka zarządzania*. Towarzystwo Naukowe Organizacji i Kierownictwa"Dom Organizatora".

Sudoł S., Kożuch B. (2010), *Rozszerzyć nauki o zarządzaniu o zarządzanie publiczne jako ich subdyscyplinę*,(In:) Ewolucja nauk o zarządzaniu – osiągnięcia i perspektywy, S. Lachiewicz, B. Nogalski, (Eds.) Wolters Kluwer, Warszawa.

Sułkowski, Ł. (2012), *Meta-paradigmatic cognitive perspective in management studies.*

Sułkowski, Ł. (2012), *Epistemologia i metodologia zarządzania*, PWE, Warszawa.

Sułkowski, Ł. (2004), Neopozytywistyczna mitologia w nauce o zarządzaniu, *Organizacja i kierowanie*, No. 1 (115).

Professor Barbara Kożuch, PhD
Mateusz Lewandowski, PhD
The Jagiellonian University
Institute of Public Affairs

Contemporary organizational humanism – overview

Abstract: Humanistic aspects of management have been investigated for decades. Their role is increasing due to their imbalance with the economic context of management. The chapter outlines the philosophical aspects, multi-sector relevance, paradigms and contemporary challenges of organizational humanism, and addresses the directions for future investigations and implementations.

Keywords: organizational humanism, philosophical aspects, trends, paradigms, challenges

One of the distinctive traits underlying the evolution of management science is its growing penetration by humanistic content. An interest in humanistic aspects of management dates back to 1930s. In essence, it emerged as a result of dissatisfaction with the existing methods of scientific work organization and classical organization theory. Since then, an orientation on people has been present – though to a varied degree – in the theory and practice of organization and management. However, for the most part, management issues connected to humanistic elements were constrained to a single field in the company's operations. These issues were principally investigated by researchers and specialists concerned with interpersonal relations, the so-called HR persons. Only the recent twenty five years has seen the development of concepts concerned with enterprise management which perceived individuals within an organization in the context of the whole system and its subsystems.

Quite independently, but more or less at the same time, there emerged a need for managers' work across public and social organizations. Since central to this kind of organization is the fulfilment of higher needs across present-day societies with organizations operating in multiple spheres of social life, organizational humanism needs to be their natural attribute. Thus, by definition, humanistic elements prevail in the management of these organizations.

The paper attempts to characterize the present-day, organizational humanism, particularly through:

- revealing some of its philosophical aspects;
- reflecting on it by using concepts concerned with the management of enterprises, public and civil organizations as well as in management trends and schools;
- linking its essence with the paradigms behind management sciences;
- indicating the most important contemporary challenges.

Analyses were conducted based on the theory of organization and today's management concepts. The authors applied the results from previous studies investigating humanistic content within an organization and management. The analyses also used philosophical studies and research on the paradigms behind social sciences in order provide insights into the analysed issues.

Philosophical aspects of organizational humanism

In order to understand the essential specifics of organizational humanism it is pivotal to demonstrate it from the perspective of philosophical issues. The connection between philosophy and organizational humanism appears to be two-fold. First, philosophical problems related to a human being tend to be broader than the problems tackled by management sciences, and they appear to be primeval, because their frame of reference goes beyond the organizational reality. Second, management sciences make use of some philosophical concepts to explore in-depth – and add to – the body of management knowledge.

A man's functioning in an organization affects spiritual, religious and family spheres, perception of oneself as a human being, experience of one's own existence, one's manner of living, perception, creativity and experience of one's relationships with others (relationships between I and You), etc. These aspects are all part of issues addressed in contemporary philosophical trends, e.g. philosophy of spirit, philosophy of existence or philosophy of life (Gadacz 2009; 2010). Specifically, ethical issues examined in the view of philosophy bear an immense importance for organizational humanism (Corlett, 1988a, 1988b). It was proved that individuals within an organization act in a spontaneous manner, take decisions, operate in chaos and under volatile circumstances, mostly propelled by feelings and emotions. This implies a gap between organizations resting on the rationality of individual and the irrational quality of human behaviour. All in all, the conclusion may be drawn that there is a need to discern individual morality, authenticity, self-realization, that is, attributes of the human side of enterprise (Denhardt & Denhardt, 2003, p. 73; Argyris, 1973, p. 261). On top of that, it is necessary to clarify which behaviour is or is not rational. And again, philosophy is a primary field where this problem may be addressed. It is not sufficient to assume that it is all about "the prism of which goals and whose values we will evaluate rationality" (Simon, 2007, p. 108). There are other rationalities; their consequence is to act impartially, not to accord any particular privileges to one's interests, or to act in such a way as to achieve the ultimate and true good of human beings (MacIntyire, 2007, p. 49).

Management sciences and the managerial practice also utilize philosophical concepts which prove to be important tools for studying and advancing organizational humanism. Research methods used in management sciences and embedded in philosophy include, for instance, phenomenology (Bombała, 2014). Other tools rooted in philosophical reflection and developing organizational humanism in practice involve the program grounded on the Rule of Benedict, which seeks to develop spiritual aspects of the man functioning in the workplace (Bianchi, 2009).

Understanding the human nature of the participants involved in organizational processes has a significant impact upon the development of organizational humanism. Humanistic science rightly highlights (Adams, 2005, pp. 34–35) that people are free creatures undertaking organized actions. Principally, they are complex entities and thus it is not only their particular attributes as physicality, behaviours, cognitive capacity instincts, human relationships and their culture that matter, but it is also a sum of these attributes. Their examination and comprehension require a reference to philosophy which provides a theoretical foundation for scientific reflection as well as expands a tool set for managers.

Fundamental assumptions of present-day organizational humanism by management schools and trends

Today's knowledge on organized activities by people was shaped as a result of the growing output produced by various disciplines. Some concepts were negatively verified by management practice while others were modified. A substantial part retained its cognitive and practical relevance despite the passage of time and immense changes in the field.

Those subscribing to the scientific work organization are focused on formulating instructions, determining timeframes, quality control, training workers and work discipline. From among fourteen principles coined by H. Fayol, there are four that brought a spotlight on human beings. These include: stability of tenure of personnel, equity, initiative and esprit de corps. In light of the classical approach, a man was treated instrumentally based on the stereotype of the economic man (*homo economicus*), a passive executor of tasks who should be motivated by material incentives.

Putting humanistic elements into management, including the concept of the social man (*homo socius*), characterizes a behavioural trend. The concept holds that interpersonal relationships tend to be a very strong source of motivation for persons' behaviours, who strive to attain recognition of their own value in their work environment. They are not only guided by material imperatives, but also by feelings and emotions (de Dreu et. al. 2004). Due to a modified concept of the social man, this trend modelled a stereotype of the man who finds self-fulfilment through work, among other things, as an active participant within an organization. The term organizational humanism emerged in the research agenda of the Human Relations School. One may list the following features of this behavioural school: conducting research focused on the satisfaction with work in relation to organizational effectiveness, on managing people, on decision-making processes and communication among persons across organizations, on conflicts and shifts in organizations, on implications of interpersonal problems, on social relationships and impact of people, on other organizational components, relations among groups, management styles, influence on formal structures, on human behaviour, etc. This trend is being continued today (Fleetwood, 2007; Mc Ewan, 2001).

The emergence of organizational humanism was thus a response to the view that persons within an organizations are propelled by an urge to attain material benefits when making their decisions.

Enquiries undermining the fundamental presumptions underlying the *homo economicus* concept were carried out by Ch. Argyris. These include, among other things, conclusions about the shifts occurring in attitudes, approaches and behaviours among employees because of their organizational maturity (Table 1).

These conclusions that transpire to be still valid and current suggest that at least some employees may offer maturity, autonomy, activity, organizational consciousness and other virtues helpful to accomplish the objectives set by an organization. Meanwhile, often by definition, and contrary to managers' declarations at times, they are expected to be obedient, submissive and limited to their work. A humanistic approach to management allowed Ch. Argyris to uncover various constraints in the rational model of organizational behaviour (Argyris 1965, 1973). In this model, it is distinctive for a management staff to determine their organizational goals and tasks to be accomplished as well as manners for rewarding and disciplining and trainings within a power hierarchy.

Table 1. The direction of shifts in organizational attitudes and behaviours

No.	State A	Shift direction	State B
1.	Immaturity		maturity
2.	Passivity		activity
3.	Dependence		independence
4.	narrow scope of operations		substantial scope of operations
5.	superficial interest		deep interest
6.	shorter perspective		long perspective
7.	position pf subordinate		position of partner or superior
8.	lesser organizational awareness		bigger organizational awareness

Source: based on: Argyris, 1957; Denhardt & Denhardt, 2003.

The systemic school distinguished, within an organization, subsystems of goals and – what is important in the context of our enquiries – values and a psychosocial subsystem. One of the few core attributes of the management sub-system is to establish a system of managing employees and their work, including motivating them.

A systemic approach introduced a stereotype of the rational man. According to this concept, people continue to make choices while taking decisions based on the profit and loss account containing material as well as non-material elements. They have the capacity of finding rational, and even optimal, solutions to new problems.

Even a superficial analysis of management schools and trends shows the evolution towards informing management with humanistic content. This issue achieved great prominence in non-business organizations where profit or – to put it differently – market value does not constitute the primary criterion for organizational effectiveness.

In the evolution of management science it is easy to notice a clear departure from pure economism and a tendency to assigning an increasingly greater role to persons within an organization as social beings, and as human beings in general, thereby putting bigger spotlight on organizational humanism.

Organizational humanism in contemporary management concepts

Diverse manners of understanding management issues co-exist across the humanities. The most telling is a recent criticism of the concept of social man (*homo economicus*) and an attempt to create a model of *homo diseconomicus* (Brockway, 1991; Morawski, 2001; Witkowski, 2009).

The hypothetical *homo economicus* pursues a single idea, i.e. commitment to its material interest. Therefore, it assigns a primary significance to economic criteria: profit or loss. This focus on economic effectiveness requires a skilful use of the possessed resources. In its behaviour, it is selfish and does not attach any importance to the human aspect of production and exchange. But it needs to behave in this way to be able to accomplish the function performed by it across different economic processes (Brockway, 1991, p. 9; Morawski, 2001, pp. 20–26; Witkowski, 2009, pp. 115–132). The concept of humanistic economy or *homo diseconomicus*, in turn, is supposed to stand in opposition to such an approach. Both these concepts are to imply the demise of the economic man. However, their major disadvantage is being white and black over how they understand the surroundings. Both pure economism and pure humanism are overtly simplified constructs, especially when referred to management sciences. A definite majority of today's theories and concepts concerned with an organization and management cover both economic and humanistic elements, actually assigning varied meaning to them and identifying their diverse scope.

Polish reference literature provides the concept of humanizing management processes which, according to T. Mendel, encompasses contemporary trends in management, striving to attain conformity with the principles of humanism and giving an opportunity to develop employees' professional and personality potentials, their skills, ingenuity, independence and responsibility (Mendel, 2004, pp. 27–28). The author views these processes at manifold specialist levels, e.g. in the social policy of the organization, in joint decision-making, in the regulation and executive sphere, in management styles, in the organization of work time and work structuring as well as in other methods such as management by objectives and human resources management methods.

Organizational humanism evolves alongside the development of management sciences. Its essence continues to be unchanged. Specifically, it brings into focus persons in the organization as social and human beings, while economism in management is essentially connected with enterprises as organizations seeking to generate profit or income.

On the whole, the issue of profit continues to have a remarkable relevance for the operations run by present-day enterprises, even though it ceased to be the only objective and criterion of the effective performance in many organizations. Today's social shifts triggered the need for conducting business operations in accordance with ethical values (Walczak-Duraj, 2002, pp. 237–240).

Currently, it is assumed that an enterprise should not pursue one single objective, but it should pursue several objectives forming a unit which links, for example:

- the objectives related to the market – hitting new markets, increased share in the market;
- the objectives related to economic effectiveness – generating specified profit, rate of return of capital involved;
- the financial objectives – maintaining financial liquidity, creditworthiness;
- the social objectives – adequate remunerations and labour conditions;
- the objectives related to surrounding setting – image in society, winning political and social influence, etc.

For example, keeping in mind the objectives of present-day small and medium-sized enterprises, it should be underlined that the most typical legal and organizational form is a company. Thus, the primary target of contemporary companies is to maximize a return on the capital invested by their owners. Numerous proprietors – particularly family-owned companies – do not recognize that profit is the principal motive while taking organized actions. Their key goal is to generate income at a level that ensures retaining ownership and ownership supervision and to achieve satisfaction with holding an enterprise and managing it (Piasecki, 2001, p. 36). Nevertheless, none of these goals may be delivered without generating profits from the conducted business operations.

Therefore, in this context it may be concluded that, even though the objectives of the enterprise evolve, they cannot be contradictory to the logic behind market processes, which prioritizes the economic objectives. Without their attainment, no other objectives may be accomplished because the profit is the source of their funding. The change of that logic is not affected by the ability to carry out specific projects thanks to the grants obtained from European funds. Thus, while economism is an inherent feature of corporate management, modern conditions require that humanistic aspects should complement it.

Present-day organizational humanism may be illustrated by the concept of the value for personnel, also labelled as an orientation on man, especially on such factors as: relationships between employees and management staff, interests in work, employment security, organizational image. A high level of humanized management processes occurs also in participative management (Germonprez & Warner, 2013; Stocki, Prokopowicz & Żmuda, 2008).

Organizational humanism also encompasses the concept of organizational development – OD (Denhardt & Denhardt, 2003, p. 37; Kożuch, 2007, pp. 241–243; Beckhard, 1969; Kegan, 1971). The OD concept is related to the strategies and culturally-oriented interventions involving a shift in values, customs, standards of conduct, attitudes and behaviours among people with the goal of establishing a new organizational culture. The proponents of this concept claim that people have a natural desire for personal improvement and development, and that employees are characterized by a strong need to be accepted by other members of the organization. At the same time, the climate across the most of groups and organizations does not facilitate an open expression of feelings. One of the OD methods, long used, but only recently discovered in Poland, is research during action, also called action research, action inquiry or organizational learning.

In recent years, there has been an increasing interest in the concept of the organization which serves its surrounding environment, and of the corporate social responsibility (CSR). Both these concepts explicitly contain humanistic elements. It is unlikely to support the position manifested by Friedman (1962, pp. 133–136), who argues that the idea of corporate responsibility only refers to resource management and generating profit in a legal way. Even though there are different approaches to CSR (Bowie, 1979; French, 1979, 1984; Fitch, 1976), and the concept itself evolved (Wartick & Cochran, 1985; Wood, 1991), a broad understanding of this responsibility prevails today. In general, it is understood in terms of running a reliable and fair business (towards clients, employees and business partners). Other important features of this model are: compliance with law, providing employees with opportunities of their development and respecting their rights, and philanthropic activities for the benefit of the community in which the company operates (Caroll, 1991; *Report...*, 2003, p. 7). Moreover, an uneven distribution of accents, including placing a greater emphasis on economic, ethical or legal aspects of the organization, is also underlined (Schwartz & Carroll, 2003). Since researchers and market processes participants differently define the key areas of corporate responsibility, it may be claimed that the gist of CSR involves formulating and attaining a range of mutually related enterprise's objectives, that is, economic, social, ecological and ethical objectives.

From the perspective of the accomplishment of the management controlling function, instruments taking into account organizational humanism are also utilized. There are numerous concepts concerned with the efficiency of management on which control systems are founded, and which provide guidelines for efficiency assessment criteria. In this regard, both purposeful and systemic concepts appear to be insufficient, and the only likely solution is to use the multi-criteria concept incorporating the needs and interests of multiple stakeholders' groups (Hall 1999, p. 253; Connolly, Conlon & Deutsch, 1980; Bielski 1996, pp. 104–123, Kulikowska-Pawlak, 2010; Lewandowski, 2012, 2013). It is essential for organizational humanism to formulate the principles for efficiency, which are based on a new logic, and which take into consideration the stakeholders' needs. Basically, the needs of broadly understood stakeholders should be considered, but this has to be based on a guarantee that obligations towards other persons and parties are fulfilled. These obligations, however, do not follow from the managers' intent or choice, but from the role he or she performs towards other persons and organizations. These assumptions are illustrated by the concept of *agathos* cited by MacIntyre – to be *agathos* means "to do what my role requires, to do it well, deploying the skills necessary to discharge what someone in that role owes to others" (MacIntyire, 2007, p. 63). The essence of effective management is the responsibility towards employees, proprietors, community, law, environment, economy, public policies and other organizations (Lewandowski, 2011; Schwartz & Caroll, 2003). To a greater extent, humanism principally occurs in management theories and concepts across public and civic organizations. The first are established to facilitate the attainment of higher objectives, thereby determining the achievers among the citizens' community. The actions undertaken in civic organizations, in turn, stem from the unsatisfied needs of people as human and social beings.

Organizational humanism in light of paradigms behind social sciences in management

Another important realm, in which organizational humanism may be characterized, is defined by the paradigms guiding social sciences that are present in management sciences. One of the most commonly adopted paradigm classification identified: the functionalist paradigm, the symbolic-interpretive paradigm, the radical structuralist paradigm and the radical humanist paradigm (Burrel and Morgan, 1979). Is there any place for an organizational paradigm in this paradigm scheme? The answer to the question should be referred to two crucial questions raised by Burrel and Morgan while designing their ideal-type paradigm matrix scheme: What is the nature of social sciences? What is the nature of society?

The primary assumptions in these two areas indicate the place of organizational humanism. Among the assumptions behind the nature of social science related to ontology, epistemology, human nature and methodology (Burell & Morgan, 1979, pp. 1–9), the assumption about the human nature has significant implications for organizational humanism. Polarized viewpoints subscribe to determinism or voluntarism. Determinism regards a man as a being responding to the environment in a mechanical manner, whereas voluntarism sees a man within an organization as a person possessing free will, being creative and shaping his or her own environment. And this is the assumption which at the level of methodology opens up opportunities for organizational humanism to become an object of research. Nevertheless, the assumption behind voluntarism is determined by some previous assumptions regarding the ontological and epistemological ground (nominalism and anti-positivism). These assumptions also influence the selection of research methods. This is justified by, for instance, the application of phenomenology in management (Bombała, 2014), the grounded theory, ethnography (Kostera, 2003) or metaphor (Morgan, 1997). These methods seem adequate to explore organizational humanism, though they are not the only ones. Yet, when taking into account the nature of society, Burrel and Morgan (1979, pp. 10–19) suggested two polarized ideal-type stances in sociology – the sociology of regulation and the sociology of radical change. For both stances, a man is important, though in different ways. The sociology of regulation regards a man as part of a community. This community operates in conformity with rules accepted by the majority of individuals, tends to be solidary, strives to be integrated and to fulfil the group's needs. In this sense, this position emphazises the subordination of the individual to the community, which gives rise to positive and negative effects for a single man. On the one hand, it may fulfil the need of affiliation and enable one to forge good relationships with others. On the other hand, the submission of the individual to the community puts domination tools into operation, thereby depriving the individual of possibilities of satisfying some needs. In this context, the structural conflict between an individual and a group as well as the urge for emancipation proves to be natural. Both stances, though contradictory, place a high premium on diverse elements of organizational humanism. Taken together, the enquiries in question reveal a crucial attribute of organizational humanism which is antinomy.

Organizational humanism is also deeply rooted in the assumptions behind critical management studies, such as: denaturalization, anti-performativity, reflexivity and emancipation (Alvesson, Willmott & Bridgman, 2009; Sułkowski, 2013; Zawadzki, 2014). Denaturalization, as a research strategy seeking to challenge the elements of organizational reality commonly regarded as natural, provides opportunities for searching and analysing various humanistic aspects. It also leads to a recognition of anti-performativity which defies social relations to be perceived only instrumentally and analysed through the prism of results maximization.

In this view, critical studies clearly demand the humanization of management. Reflexivity entails, among other things, a necessity of adopting certain values at each phase of the research process, and therefore a researcher is not a neutral observer and refers implicitly or explicitly to organizational humanism values, either embracing them or rejecting in whole or in part, which thus translates into an interpretation of research findings in line with the logic behind the hermeneutic circle. This supports the fact that researchers examining economic aspects in management mostly ignore humanistic aspects and vice versa. The last assumption is a quest for emancipation, which in this context entails change in the established organizational order. Accordingly, emancipation translates into the accomplishment of organizational humanism ideals in management practice.

These paradigms underlying management sciences and the development of management concepts show the challenges to organizational humanism.

Challenges to contemporary organizational humanism

The present-day managerial practice and scientific research in the field of organizational humanism are faced with serious difficulties. These challenges concern, among other things, the economization trap, a wrong research methodology or a lack of development model.

1. The economization trap means that management concepts and practices increasingly incorporate a humanistic aspect of management which at the same time may be a propelling driver for an even more efficient economization, specifically across business organizations. The profit is critical. This gives rise to a dilemma whether humanistic concepts serve against their ideals to better steer employees' behaviour in the economic interest rather than emancipation interest. After all, what should an organization look like where a man is given primacy rather than profit or another target? Should a man be more important in organizations than their objectives and expansion?
2. Organizational humanism turns a researcher's attention to barely perceptible and intangible aspects of man's functioning within an organization. This presents methodological problems and difficulties with conferring adequate status to scientific knowledge. Different studies on organizational humanism do not fit within positivistic science ideals. Not all scientific circles involved in management sciences are unwilling to agree with it. Hence, distinct methodologies used for investigating organizational humanism are its weakness for some circles, and the knowledge obtained though the "understanding" methods is depreciated. Furthermore, against the backdrop of research on organizational

humanism, an effect of hermeneutic circle may be recognized – researchers of economic and humanistic aspects rarely analyse these aspects in a systemic manner.
3. Organizational humanism seems to be insufficiently developed in the practice of business, public and civic organizations. This is, in fact, an intuitive statement. Since there is no organizational humanism model allowing for an assessment of its advancement degree within a specific organization. However, it seems that the nature of management science gradually provides appropriate foundations in the form of the cumulative nature of science, the development of the concept of management, the present-day paradigmatic reflection and the acceptance of methodological pluralism.

This list fails to thoroughly pinpoint the challenges facing organizational humanism. However, the challenges outlined above appear to be not only critical but also urgent.

Conclusions

Enquiries make it possible to formulate the following assertions characterising present-day organizational humanism.
1. Change in the organizational environment increasingly creates the demand for humanized management.
2. Contemporary organizational humanism evolves regardless of the organization type. However, it cannot replace economism which is connected with the logic behind market processes.
3. The promotion of knowledge about the essence of organizational humanism may contribute to its more frequent implementation into management of contemporary organizations.
4. Greater awareness of paradigm distinctions in management science should reduce the depreciation of a research methodology deployed for examination of humanistic management aspects. This facilitates systemic research of economic and humanistic aspects, which is required because of the antinomy of organizational humanism and the economization trap.
5. An in-depth understanding of the essence of organizational humanism requires a reference to management philosophy.
6. The primary challenges for organizational humanism relate to the manner of shaping relations within an organization, so as to minimize the effect of the economization trap, establishing a model for the development of organizational humanism and promoting anti-positivist science ideals.

References

Action Science, What Is Action Science? http://www.actionscience.com/actinq.htm#theory (Accessed 11.02.2009).

Adams A. A. (2005), *What Does It All Mean?* Imprint Academic, Exeter.

Alvesson M., Willmott H., Bridgman T. (2009), The *Oxford Handbook of Critical Management Studies.* Oxford University Press, Oxford.

Argyris C. (1957), *Personality and Organization*, Harper Collins, New York.

Argyris Ch. (1965), *Zrozumienie zachowania ludzkiego w organizacji: jeden punkt widzenia*, (In:) Nowoczesna teoria organizacji, M.Haire (Ed.), PWN Warszawa.

Argyris C. (1973), Some Limits of Rational Men Organization Theory, *Public Administration Review*, No. 33.

Argyris C. (1987), *Reasoning, action strategies, and defensive routines: The case of OD practitioners*, (In:) R. A. Woodman & A.A. Pasmore (Eds.), Research in organizational change and development. Volume 1, JAI Press, Greenwich.

Beckhard, R. (1969), *Organization Development: Strategies and Models*, Addison-Wesley, Reading, Mass.

Bianchi P.G. (2009), *Duchowość i zarządzanie*, Tyniec Wydawnictwo Benedyktynów, Kraków.

Bielski M. (1996), *Organizacje – istota, struktury, procesy*, Wydawnictwo Uniwersytetu Łodzkiego, Łódź.

Bombała B. (2014), Phenomenology as the epistemological and methodological basis of management sciences, *International Journal of Contemporary Management*, 13(1), 150–172.

Bowie N.E. (1979), *'Changing the Rules', in Ethical Theory and Business*, T. L. Beauchamp and N.E. Bowie (Eds.), Englewood Cliffs: Prentice-Hall, pp. 147–150.

Brockway G.P. (1991), *The End of Economic Man*, W.W. Norton & Company, New York, London.

Burell G., Morgan G. (1979), *Sociological Paradigms and Organizational Analysis: Elements of the Sociology of Corporate Life*, Ashgate Publishing, Burlington.

Carroll A.B. (1991), The Pyramid of Corporate Social Responsibility: Toward the Moral Management of Organizational Stakeholders, *Business Horizons* (July-August 1991), pp. 39–48.

Connolly T., Conlon E.J., Deutsch S.J. (1980), Organizational Effectiveness: A Multiple-Constituency Approach, *Academy of Management Review*, Vol. 5, No. 2, pp. 211–217.

Corlett J.A. (1988a), Schefflerian Ethics and Corporate Social Responsibility, *Journal of Business Ethics*, Vol. 7, No. 8 (Aug.), pp. 631–638.

Corlett J.A. (1988b), Corporate Responsibility and Punishment, *Public Affairs Quarterly*, Vol. 2, No. 1 (Jan.), pp. 1–16.

Denhardt J.V. i Denhardt R.B. (2003), *The New Public Service. Serving, not Steering*, M.E. Sharpe, New York, Armonk.

de Dreu C., West M., Fischer A., MacCurtain S. (2004), *Origins and consequences of emotions in organizational teams*, (In:) R.L. Payne, L.C. Cooper (Eds.), Emotions in Work: theory, research and applications for management, John Wiley & Sons, Chichester.

Fitch H.G. (1976), Achieving Corporate Social Responsibility, *The Academy of Management Review*, Vol. 1, No. 1 (Jan.), pp. 38–46.

Fleetwood S. (2007), Serching for the human in empirical research on the HRM-organizational performance link: a meta-theoretical approach,(In:) S.C. Bolton, M. Houlihan (Eds), Seraching for the human in human resource management, Palgrave Macmillan, Houndmills.

French P.A. (1984), *Collective and Corporate Responsibility*, Columbia University Press, New York, pp. 31–47.

French P.A. (1979), The Corporation as a Moral Person, *American Philosophical Quarterly* 16, 207–215.

Friedman M. (1962), *Freedom to Choose*. University of Chicago Press, Chicago.

Gadacz T. (2009), *Historia filozofii XX wieku*. Vol. 1 & 2. Znak, Kraków.

Germonprez M., Warner B. (2013), *Organizational Participation in Open Innovation*, (In:) J.S.Z. Eriksson Lundström, M. Wiberg, S. Hrastinski, M. Edenius, P.J. Ågerfalk (Eds.), Managing Open Innovation Technologies, Springer, Wien.

Hall R.H. (1999), *Organizations – sutructures, processes and outcomes*, Prentice Hall.

Kegan D.L. (1971), Organizational Development: Description, Issues, and Some Research Results, *The Academy of Management Journal*, Vol. 14, No. 4 (Dec.), pp. 453–464.

Kostera M. (2003), *Antropologia organizacji. Metodologia badań terenowych*, PWN, Warszawa.

Kożuch B. (2007), *Nauka o organizacji*, CeDeWu, Warszawa.

Kulikowska-Pawlak, M. (2010), *Pojmowanie efektywności organizacji – definiowanie, pomiar*, (In:) Pomiar efektywności organizacji publicznych na przykładzie ochrony zdrowia, A. Frączkiewicz-Wronka (ed.). Wydawnictwo Akademii Ekonomicznej w Katowicach, Katowice.

Lewandowski, M. (2011), Sprawność zarządzania z perspektywy humanistycznej, *Współczesne Zarządzanie*, No. 1, pp. 106–115.

Lewandowski, M. (2013), *Introduction to Academic Entrepreneurship*,(In:) A. Szopa, W. Karwowski & P. Ordóñez de Pablos (Eds.), Academic Entrepreneurship and Technological Innovation: A Business Management Perspective (pp. 1-28). Information Science Reference, Hershey, PA,.

Lewandowski, M. (2013), *Zarządzanie strategiczne w instytucjach kultury*, Wydawnictwo Con Arte, Katowice.

MacIntyre A. (2007), *Czyja sprawiedliwość? Jaka racjonalność?* Wydawnictwa Akademickie i Profesjonalne, Warszawa.

McEwan T. (2001), *Managing Values and Beliefs in Organizations*, Prentice Hall, Harlow.

Morawski W. (2001), *Socjologia ekonomiczna. Problemy, teoria, empiria*, PWN, Warszawa.

Morgan G. (1986), *Images of organization*, Sage Publictions, Newbury Park – London – New Delhi.

Mendel T. (2004), Humanizacja procesów zarządzania w organizacjach XXI wieku, *Zeszyty Naukowe Akademii Ekonomicznej w Poznaniu*, 48/2004.

Owen H. (2000), *The Power of Spirit. How Organizations transform*, Berret-Koehler Publishers, San Francisco.

Piasecki B. (2001), *Ekonomika i zarządzanie małą firmą*. PWN, Warszawa,

Raport z badań *"postawy wobec społecznej odpowiedzialności biznesu w Polsce"* (2003), Fundacja Komunikacji Społecznej, Warszawa.

Schwartz M.S., Carroll A.B. (2003): Corporate Social Responsibility: A Three-Domain Approach, *Business Ethics Quarterly*, Vol. 13, No. 4 (Oct.), pp. 503–530.

Simon H.A. (2007): *Podejmowanie decyzji i zarządzanie ludźmi w biznesie i administracji*, Wydawnictwo Helion, Gliwice.

Stocki R., Prokopowicz P., Żmuda G. (2008): *Pełna partycypacja w zarządzaniu*, Wolters Kluwer Polska, Kraków.

Sułkowski Ł. (2013), Paradygmaty nauk o zarządzaniu, *Współczesne Zarządzanie*, 3/2013.

Walczak-Duraj D.(2002), *Ład etyczny w gospodarce rynkowej: doświadczenia polskiej transformacji*, Wydawnictwo Uniwersytetu Łódzkiego, Łódź, pp. 237–240.

Wartick, S. L., and P. L. Cochran. 1985. The Evolution of the Corporate Social Performance Model. *Academy of Management Review* 10(4), pp. 758–769.

Witkowski L. (2009), *Jak pokonać homo economicus?*(In:): Humanistyka i zarządzanie, P. Górski (Ed.), WUJ, Kraków.

Wood, D.J. (1991), Corporate Social Performance Revisited, *Academy of Management Review 16(4)*, pp. 691–718.

Zawadzki M. (2014), *Nurt krytyczny w zarządzaniu. Kultura, edukacja, teoria*. Sedno Wydawnictwo Akademickie, Warszawa.

Part III
Contemporary Management Dilemmas

Colonel Marek Bodziany, PhD
Tadeusz Kościuszko Land Forces Military Academy

Paweł Kocoń, PhD
University of Economics

Military science and management science – methodological connection in the context of culture of organization

Abstract: This chapter describes the similarities and differences between management and military science. The description is focused on the notion that is strategic for management, namely – the organization culture. Military science shares many common elements with management science. It is the organization culture that may decide about both the performance of an organization and the result of a military conflict.

Keywords: management science, military science, organizational culture, methodology, army

Introduction

Military science and management science are widely connected – both discuss human activities. Dealing with management science (especially from the humanistic perspective), one may explain the issues associated with warfare through the perspective of planning, organizing, deciding, motivating and controlling. Military science, in turn, is the source of inspiration for managers; this seems to be proved by the comments and supplements offered to the classic book dealing with the strategy: *The Art of War* written by Sun Tzu (Garinaldi & Sun Tzu, 2005).

An analysis of the connection between military science and management science requires theoretic reflection, the cognitive sense of which is due to the dynamics of changes that took place in science and scientific disciplines classifications. As a result of those changes, there appeared two scientific disciplines within social sciences, called security studies and military science (Journal of Laws, 2011, No. 179, Item 1065). Security sciences were created as an effect of the decomposition scientific areas, previously attributed to military science only. Although the ideas to separate security studies as an autonomic, new scientific area appeared in Western science in the middle 1990s, in Poland this discipline was created only in 2011 (Baldwin, 1995).

Therefore, the question arises: what factors influenced its establishment? The answer seems to be difficult and easy at the same time. Such a dual approach is based upon two cognitive pillars, where the change of scientific perception of security is, on the one hand, due to the dynamics of social-political processes, which took place after the bipolar line dividing the world had been destroyed and the new threats to safety appeared, and on the other hand – due to the shift in the paradigm of world peace.

It is due to the nature of contemporary asymmetric military conflicts that the key to the permanent peace is, among other things, the reconstruction of a country, meaning the military forces implementing the tools of public management. The crucial factor for the victory in an asymmetric conflict is the social acceptation of military actions – thus, the armed forces started using public relations techniques (Kryszk, 2007). The basic tasks for the armed forces is being the pillar of national defence and power, in which natural resources, demographic and geographic potential, social morale, the quality of governance and diplomacy and the military potential of a country are included (Morgenthau, 1973).

Moreover, the armed forces fulfil many other roles in contemporary democratic society. Less and less often they are a direct defender, which is due to the fact that military conflicts – especially those highly intensive – take place in authoritarian or totalitarian countries mostly. Armed forces in democratic societies have an extremely important role, being an institution socializing to the roles of a citizen, employer and a contractor, and thus in some ways – to the role of a collective citizen; they are an institution using regional resources and, at the same time, through direct and indirect taxes payment, the animator of an economic growth.

The authors' main interests were focused on the issue of organization culture. In the article, the authors aim to discuss issues such as how the idea of organization and climate culture is visible in highly specific organizations like military units. To that end, a sharp focus will be placed upon the relations of organization culture and organization climate in the military unit and upon its morale. The authors will also examine the way in which the organization culture of the military unit influences its military value and its actions in crisis management, which, in turn, is strictly connected with public management.

Management science and military science- definitions and differences

In order to sketch the demarcation lines and common areas for various fields of life, one ought to define those areas. Defining particular social sciences, such as management science or military science, allows one to find both commonalities and differences between their particular elements. According to R. Griffin, management is "the set of actions (including planning and decision making, organizing, leading, managing people), directed to the resources of the organization (human, financial, fixed and information) and performed in order to reach the company goals in effective and efficient way" (Grifiin, 2005, p. 6). Griffin limits the understanding of the issue of managing to the area of an organization. An organization is defined by T. Kotarbiński as "certain type of an entity in which all parts play a role in the entity's success" (Kotarbiński, 1958 p. 75). Thus, an organization is not only a subject examined by management science, but also by other scientific disciplines, especially social sciences. T. Pszczołowski does not limit management to the organization. He believes that management is "an activity based on disposing the resources; as people belong to the most important resources, money also belong to resources, and through it one influences people. Managing is related to directing. We tend to link the terms organization and management, directing and management" (Pszczołowski, 1978, p. 288).

Management processes are based on four functions: planning, organizing, directing, and controlling (Machaczka, 2001, pp. 36–37). S. Sudoł presents management science as such, which "is a socially useful knowledge in the form of a set regulations of economic or social life and the theory which explains the specific area of reality and/or helps rationalize this reality" (Sudoł, 2007, p. 8). The vastness of knowledge needed to learn the mentioned social and economic life makes, as Jerzy Niemczyk writes, management science to be the collection of various disciplines, sub-disciplines and scientific specializations (Niemczyk, 2011, p. 32). That is the reason why, as Jerzy Niemczyk states, "management is polimethodological, which means that its methodology is rich and various. It covers both cognitive and pragmatic methods taken from other scientific disciplines" (Niemczyk, 2011, p. 35). Such polimethodology has nonetheless a huge flaw. It is difficult to identify paradigms, methods and techniques proper for management science, which will differentiate it from, for example, the sociology of organization. Ł. Sułkowski is right to observe that it is difficult to clearly point to a crystallized subject of management science, which would be distinctly different from other scientific branches (Sułkowski, 2012, p. 61). Summing up, one ought to highlight that management science is crucial to understand the activities of not only commercial but also public organizations. The latter were examined in a sub-discipline of management, which is public management.

The military science – the range and the methodology

The starting point to analyse the methodology of military science is the issue of security. The problem of security is constantly deepening and incorporates a new range of contents. As J. Stefanowicz claims, the modern, wide-ranged understanding of security is based on taking into consideration all potential threats to our material or moral existence, wealth and the development of the individuals, social groups, nations and humanity in general; military threats are only one type of such dangers, apart from economic, ecological, political and ideological ones (Stefanowicz, 1993, pp. 10–11). Therefore, based on the thought described above, one may assume that the contemporary idea of security is the awareness of the threats that endanger different elements of national power, some of which were listed above.

Whereas the cognitive sources have been quite thoroughly described in the literature (Koziej 2010, Wołejszo & Jakubczak, 2014), there are still many gaps concerning the methodology of examination of this scientific discipline. Based on them, there appeared a lot of dilemmas, due to the fragmentation of the above mentioned branches of knowledge which create military science, as each of them has a well-established methodology, in many cases specific and varied. This shows that the methodology of military science itself constitutes a serious challenge due to the need to find a methodological common denominator. It seems that this dilemma will not be solved soon, and the trials to create a well-based methodology now do not give the required effect. The difficulties derive from the impossibility of universalizing methodological approaches based on various disciplines, e.g. sociology, empirical research (qualitative and quantitative) or history with its examination of archives.

This is just part of the reality with which the methodology of military science clashes. One ought to realize that although they are situated in social sciences, in many cases their methodology does not apply to the needs of examining the specific notions. It often requires a triangulation of the methods applicable to various disciplines and the intuition to choose those methods well, which also concerns the order in which they are used. Such a triangulation means nothing more or less than that various methods should be applied in order to address the needs of examining a specific notion or a process.

Nonetheless, it is worth highlighting that despite the methodological particularities, the majority of sciences co-creating military science rely upon some common methodological paradigms, which are universal and utilitarian for all. It should be noted that the methodology of security study always concerns two zones (or dimensions): the social nature of security (meaning its ontology) and our knowledge about it (meaning epistemology). The ontological dimension concerns the nature of reality, that is, the question of security of various objects. One can present this nature on the objectivity-subjectivity axis, therefore trying to answer the question whether it is an objective reality that exists in the external world, or merely a subjective creation of people (politicians, observers, common people, researchers). The second, epistemological dimension concerns methods of gaining the knowledge about security of individuals. Thus, we have two approaches, one of which makes it possible to explain social phenomena (including security), and the other makes it possible to understand them by interpretation (Zięba, 2012, p. 17).

Based on methodological foundations of social sciences, it is justified to divide the methodology of security studies into positivistic (empiric) and post-positivistic. The first group contains: the classic approach (traditionalism, derived from history, philosophy and law), realism (neorealism), liberalism and behaviourism. The post-positivistic approach, in turn, is related to critical theory, postmodernism and constructivism (Zięba, 2012). The most effective methodology of examining security in the conditions of civilization changes seems to be the transition, suggested by Max Weber, from interpreting and understanding to explaining the issue of security of various objects. In general, the subject of research in military science will always be any processes and phenomena – as well as the relations among the elements of structures taking part in these processes – that are situated in the social context.

Military science is defined in a variety of ways. For example, military science is defined as a systemized body of knowledge related to the theory of application and employment of military units and weapons in land warfare and armed conflict. In that sense, it encompasses issues related to the following areas: military leadership, military organization, military training and education, military history, military ethics, military doctrine, military tactics, operations and strategy, military geography and military technology and equipment (Piehler,2013, p. 881). It is also a theory, and practice, of preparing and running both armed and not armed operations performed by the armed forces during the times of peace, crisis and war (Zieliński, 2002). Thus, military science is linked to a specific situation of war or military conflicts. It addresses the role played in such situations by properly led, equipped, trained and highly moraled military units. Military science is defined by all activities run by the armed forces. However, there is no information concerning its proper methods and techniques of research.

Within the Dewey Decinal System, military science is classified in the social sciences class (Piehler, 2013, p. 881). It means that military science is closely related to management science, as they both include similar parts, such as management. The question of leading ought to be enumerated as the first one. The editor of a popular magazine concerning leadership, J. Kręcikij, points to the place leadership takes in management theory (Kręcikij, 2007, p. 11), by comparing various definitions of management and leading processes. He goes on to conclude that leading is a specific form of managing (Kręcikij, 2007, p. 17).

Now, we can proceed to present some examples of the common areas. Undoubtedly, the roots of the (widely understood) military is logistics. Military logistics does not differ much from civilian logistics, connected, for example, with transporting dangerous materials; moreover, military units are supplied with objects of civilian nature – food, fuel, spare parts, press, post, clothes, etc. Delivering such objects to the military unit is not different from delivering the same things to numerous groups of people – staying in a hotel, participating in a festival, etc. The major difference between civil and military logistic chains seems to be in the security system. Almost each of military logistic chains is masked and protected against the attack. There are also civil transports that are protected – e.g. of priceless pieces of art – but the scale of masking and armed protection is definitely smaller. What is more, the common outsourcing of military logistics to civil companies proves the mutual influence of military and civil logistics (Jałowiec, 2010, pp. 62–65). Military roots of public relations is underlined by Teresa Święćkowska in her book *Public Relations and Democracy* (Święćkowska, 2008).

Commenting on public relations, it is important to notice certain methodological irregularity. Military public relations are called Military Public Affairs, which can be literally understood as internal affairs of the armed forces. This understanding is in accordance with NATO documents (*Allied Command...*, 2010). Armed forces also recruit and select the candidates who want to become soldiers, thereby managing their human resources. Apart from common disciplines and sub-disciplines, management and the military use similar methods, such as observation, quantitative data analysis or document analysis.

Organization culture from the perspective of a manager and of a soldier

According to Ed Schein, organization culture "is the pattern of repetitive assumptions created, discovered or developed by the given group in the process as they learn to deal with the problems of external adaptation and internal integration. It worked well enough to be perceived as right. Therefore, it may be taught to new group members as a proper way of perceiving, thinking and feeling when related to those problems" (Schein, 1990, pp. 245–262). Thus, as A. Koźmiński claims, culture is the ideal regulator (Koźmiński, 2005, p. 181). At the same time, however, William J. McIver claims that "organization culture is the collection of social norms and values that are stimulants of the institution members' behaviour, important from the perspective of the relationships which are crucial for the realization of the specific target and which occur in time and space, among people and elements of architecture" (Pietkiewicz & Kałużny, 1993, p. 47). Here is a list of cultural indicators as conceived by L. Zbiegień-Maciąg:

- organization scenario, deriving from the scenarios of the founders, or dominant leaders;
- philosophy governing an organization policy;
- core values which define the philosophy or mission of a company;
- organization climate, work attitude, the level of personal responsibility for work;
- rules of the "game", of improvements in the organization;
- customary and traditional ways of thinking and acting.

What connects a workplace with a military unit is mostly the morale of the workers. Morale, derived from Latin *moralis*, is translated as a fighting spirit, a will to fight and to achieve victory. Morale is also understood as the willingness to fulfil one's duties and to endure difficulties and dangers. It is also a feeling of responsibility and a strong belief in success or somebody's moral approach. The *Reader's Companion to Military History* defines morale as a spiritual quality considered to be desirable in soldiers and a sublime, self-denying state giving troops a sense of purpose higher than that of individual survival (Gerras, Wong & Allen, 2008). Morale is therefore an organization quality, a prism through which one assesses the qualitative aspects of military values of military units. Similarly to other organizations, military units also have their organization culture. It is often mentioned that special companies of military areas have developed organization cultures (Gerras, Wong & Allen, 2008).

Turning to the literature concerning organizational behaviour, organizational culture appears to be a context-free version of the context-specific military culture. While military culture is often used effectively as an overarching label for the military's character, way of thinking or values, there is little literature that defines the term military culture, categorizes or delineates the values that military culture claims to capture or, more importantly, provides methods or techniques of changing the military culture (Scales, 1993, p. 359). As M. Bodziany writes, "army as a social class has its own, relatively hermetic and specific organization culture, based on a specific power system, group symbols system, language system, value system, traditions written into the history of the country and the nation, especially in military conflicts" (Bodziany, 2009).

Organization culture has to be present in the army, as the armed forces are a specific organization. On the one hand, the strong standardization of artefacts – equipment, weapons and uniform – hides the organization culture. On the other hand, the regionalism of military units or the competition between them makes their organization cultures differ from one another. Charismatic leadership is not without influence here.

It ought to be noted that not only the army but also particular types of the armed forces have their own organization cultures. Inheriting the tradition of the Home Army, on the one hand, and the Polish People's Army on the other, points to the difficult development of specific organization cultures in the standardized environment. Those cultures may differentiate the activities of those units in the time of ultimate trial.

Conclusions

Both military science and management science relate to specific resources of management. They use such measures as effectiveness and action efficiency. Military science relates directly to management science.

If every organizational activity is perceived from the perspective of planning, organizing, directing and controlling, military science seems to be a sub-discipline of management science. However, if the process of organizing is perceived from the perspective of wasting competition, creating one hegemon and destroying other market participants, management will be part of military thought.

One may state that management differs from the military in a way that it is a positive (not negative) form of cooperation, but contemporary small-scale military conflicts require multi-field cooperation with the local community, constant redefining of alliances, building a country and administration, and therefore they are based on positive cooperation.

It seems that the connector between military science and management is praxeology. As T. Kotarbiński writes, "praxeology is the science of action, a general theory of action. Praxeologists examine are interested with all activities, but only from the perspective of a good job as such, that is to say, independently from any emotional motives or secondary effects, both pleasant and unpleasant. Only technical values of an activity are important to them. What ensures the effectiveness of an activity, when the job is the most economical, what factors make the action closer to a masterpiece – those are the problems of a praxeologist" (Kotarbiński 2003).

Praxeology, as a science concerning effective action, defines this effectiveness. It is composed mainly of effectiveness and efficiency, but also profitability, which means the dominance of the usefulness of the action's result over the covered expenditures (Dębski & Dębski 2011 pp. 59-67). In other words, praxeology and management use the measures of effectiveness and efficiency. Military science also uses such measures. Nonetheless, in military science the scientists have to deal with such issues as heroism, honour or fidelity. Those are completely omitted by management. Both sciences, through mutual infiltration, give people some knowledge defined as information resources. It is, however, a knowledge organized in a particular way: it is an answer to the intentions of its creators and users (Koźmiński 2005).

For the armed forces, the relations between organizational culture and organizational climate of a military unit are the most important issues. They will also focus on the way organization culture of a military unit influences its military worth and its actions. In the context of organization culture, one ought to consider the failure of some military units and victory of others, despite the unified training and equipment or similar conflict conditions. Moreover, the effectiveness and efficiency of the actions undertaken during a crisis depend, among other things, on organization culture. The norms and the values may be only apparently unified – in reality, they may be varied thanks to the so-called hidden programs. The differentiation of organization cultures is based on territorial differences among the units. The effect of this differentiation is different military value of similar military units, which, in turn, has serious consequences in the battlefield.

References

Allied Command Operations and Allied Command Transformation (2010), Public Affairs Handbook.

Baldwin D. A. (1995), *Security Studies and the End of Cold War*, "World Politics", Vol. 48, No. 1, October; cf. R. Zięba, *O tożsamości nauk o bezpieczeństwie*, Zeszyty Naukowe AON, No. 1(86), 2012.

Bodziany M. (2009), Armia a nowa jakość ładu społecznego, *Zeszyty Naukowe WSOWL*, No. 1 (151).

Dębski S., Dębski D. (1994), *Ekonomika i organizacja przedsiębiorstw*, WSPiP, cf. *Prakseologiczne aspekty prognozowania* (2011), Prace Naukowe UE Wrocław, No. 185. Wydawnictwo Uniwersytetu Ekonomicznego we Wrocławiu, Wrocław.

Gagliardi G, Sun Tzu. (2005), *Sztuka wojny. Sztuka marketingu*, Onepress, Gliwice.

Griffin R. W. (2005), *Podstawy zarządzania organizacjami*, PWN, Warszawa.

Gerras S. J, Wong L, Allen C. D (2008), *Organizational Culture: Applying A Hybrid Model to the U.S. Army*. U.S. Army War College November 2008 www.carlisle.army.mil/.../Organizational%20Culture/.

Jałowiec T. (2010), Firmy w brytyjskich siłach zbrojnych. "Przegląd Logistyczny", No. 1.

Jemioło T, (1994), *Uwarunkowania dla edukacji dla bezpieczeństwa*, (In:) *Edukacja dla bezpieczeństwa*, materiały z konferencji naukowej, R. Stępień (Ed.), AON Warszawa.

Kotarbiński, T. (2003), *Prakseologia*. Zakład Narodowy im. Ossolińskich Wrocław.

Kotarbiński T. (1958), *Traktat o dobrej robocie*. Zakład Narodowy im. Ossolińskich, Wrocław – Warszawa.

Koziej S. (2010), *Teoria Sztuki wojennej*. PISM Warszawa.

Koźmiński A. K. (2005), *Zarządzanie w warunkach niepewności Podręcznik dla zaawansowanych* PWN Warszawa.

Kręcikij J. (2007), Miejsce dowodzenia w teorii zarządzania, (In:) Podstawy Dowodzenia Kręcikij J, Wołejszo (Eds.). JAON Warszawa.

Kryszk D. (2007), *Planowanie współpracy sił zbrojnych z mediami podczas konfliktów zbrojnych (wybrane zagadnienia)*.(In:) Public relations. Teoria i praktyka komunikowania. H. Przybylski (Ed.). AE Katowice, Katowice.

Machaczka J. (2001), *Podstawy zarządzania*, AE Kraków, Kraków.

Morgenthau H, (1973), *Politics Among Nations: The Struggle for Power and Peace*, Alfred A. Knopf, New York.

Niemczyk J, (2011), *Metodologia nauk o zarządzaniu* (In:) *Podstawy metodologii badań w naukach o zarządzaniu* red W. Czakon red Wolters Kluwer Warszawa.

Obronność, teoria i praktyka (2014), ed. J. Wołejszo J. Jakubczak, Bellona Warszawa.

Piehler G. K. (2013), *Encyclopedia of Military Science* SAGE Publications, Thousand Oaks.

Pietkiewicz, E, Kałużny S. (1993), *Bankowcy i dobre obyczaje*, PWE Warszawa.

Pszczołowski T. (1978), *Mała encyklopedia prakseologii i teorii organizacji*, Ossolineum, Wrocław – Warszawa – Kraków – Gdańsk.

Scales R. H. (1993), *Certain Victory* Office of the Chief of Staff, U.S. Army Washington.

Schein E. (1990), *What is culture* (In:) Reforming Organizational Culture red Frost P. J, More L. F, Luis M. R, Lundberg C. C Martin J, Sage Publications, Newbury Park C. A.

Sudoł S. (2007), *Nauki o zarządzaniu. Węzłowe problemy i kontrowersje*, Dom Organizatora, TNOiK, Toruń.

Sułkowski Ł, (2012), *Epistemologia i metodologia zarządzania*, PWE, Warszawa.

Stefanowicz J. (1993), *Rzeczpospolitej pole bezpieczeństwa*, Warszawa.

Strategy in the Contemporary World an introduction to strategy studies (2013), J. Baylis, J. J Wirtz, Gray C. S, (Eds.) Oxford University Press, Oxford.

Święćkowska T, (2008), *Public relations a demokracja*, UW, Warszawa.

Zbiegień Maciąg L. (1999), *Kultura w organizacji. Identyfikacja kultur znanych firm*, PWE, Warszawa.

Zieliński J. (1979), *Podstawowe założenia dydaktyki sztuki operacyjnej*, AON, Warszawa.

Zięba R, (2012), *O tożsamości nauk o bezpieczeństwie*, Zeszyty Naukowe AON, No. 1(86).

Sławomir Olko, PhD
Silesian University of Technology
Organization and Management Faculty

Management of the multilateral ventures in the networks and clusters in creative industries from the perspective of the activity network theory – methodological aspects

Abstract: This chapter studies the role of the network of activities, especially for analysing and managing multilateral ventures and projects, undertaken in clusters and innovation networks. The author's methodology of analysing the activity network was described through the implementation of standards for creating an event-driven process chain (EPC).

Keywords: network of activities, clusters, multilateral ventures, creative industries, event-driven process.

Introduction

The undertaking of ventures in multilateral networks of interdependent actors poses a very important question for contemporary economics and management science. These phenomena are occurring more frequently in clusters and networks established for creating and developing new, useful products for society. Managing ventures in the networks faces a number of practical and cognitive problems related primarily to the loose nature of the relationship in the networks as well as the dynamic character of the relations. We speak deliberately of ventures, not projects that have contractors, goals and objectives, deadlines, but above all, expected results – products of the projects (PMBOK, 2012). On the other hand, ventures in the loose social and business networks (like creative clusters) are carried out based on the interconnected activities of network actors, which are agreed communication processes (Luhman, 2007). The activities lead to the projects realized by consortia of the network actors and specified resources and products.

Ventures realized in clusters and innovative networks are similar to self-organizing social systems operating according to Giddens' theory of structuration. According to Giddens' idea of structural duality, structures limit human choices, but at the same time social structures are created as a result of the choices. Giddens' theory emphasizes the importance of small changes leading to the target structure and the dynamics of social structures, especially those shaped by individuals (Giddens, 1984). In Luhmann's approach to analysing social systems, their autopoiesis (self-creation and self-regulation) is based on two elements: communication and activities. These two, closely related elements can shape the functioning of whole systems without power or coordination (Luhmann, 1995).

Innovative processes specifically refer to the phenomena of creativity and activities of creative environments. In recent years, we have observed a special interest in these environments because of the created economic and social values – a significant impact on other sectors and the social environment. For the researchers, creative clusters are an interesting proving ground for observing the phenomena of collective creativity (swarm creativity). Creative industries are focused entities creating an added value based on individual and collective creativity – the human ability to create new solutions. As A. Klasik formulates it: "In the broadest sense 'creative industries' are these sectors of activities sourced from created copyrights, patents, projects and trademarks. Creation of creative industries needs commercialisation of the above-mentioned intellectual property utilizing market and non-market values of culture and science, or more broadly: talents and knowledge, individual creativity and experimentation" (Klasik, 2008, p. 14).

The proposed methodology for analysing activity networks will be applied to the study of the so-called creative clusters – clusters existing in the creative industries. The following definition of a creative cluster will be adopted: "a group of cooperating organizations and individuals originating from local and regional communities, representing business, science, the arts, culture, education, health, entertainment and leisure activities. The dynamics of the cluster is based on the creation of a regional identity, the innovative utilization of resources and a talent search with the protection and development of local and regional values. The creative clusters are a reservoir of creative resources and skills for other clusters and innovative environments" (Knop et al., 2013).

Creative clusters and their members can undertake significantly diversified activities concerning: design, promotion (advertising and other forms of marketing communication), activities in the media (internet, television, radio, press and publishing), the activity of the ICT sector related to software creation, the widely understood activity in art and culture concerning both the creation and delivery of cultural values. According to the European Cluster Observatory (clusterobservatory.eu), the creative and cultural industries (CCI) are represented by 41 groups of the European classification of activities (NACE) describing in detail the activities of the sectors mentioned above. According to preliminary research carried out in the project, there are 30 creative clusters in Poland. Most of them are located in large cities and are connected with the ICT and multimedia sectors. Direct research enables the refinement of information about creative clusters and their activities. The obtained results will show the description of activity network of the clusters/networks – a structured and recurrent way of creating value in the cluster.

The network of activities as a cognitive perspective

In the social and management sciences, the network is considered as a set of entities (actors) connected by different kind of relationships. Network actors (nodes) represent organizations or people, while relations represent the transfer of values (products, values, knowledge) and impacts (formal interaction, political and interpersonal). Such an approach, considering the network as an organization having at least agreed boundaries and a general purpose being the result of the intentions of network actors, is represented by a large number of researchers in management sciences (de Bruijn & Heuvelhof, 2008; Czakon, 2012; Niemczyk, 2013; Stachowicz, 2011, 2014). A network approach in management sciences comes from a broader network approach in social sciences, in particular the actor-network theory (ANT). Miettinen, who conducted a critical analysis of the implementation of the actor-network theory to analyse the innovation processes, notes that mediation is the category focusing on the analyses of the prioritization of activities and the influence of the actors. Replacing a human by a technical entity having similar reactions in the network, specific for the network theory, is in fact the question about the role of the human and his or her consciousness in the network (Miettinen, 1999). The activity of actors in the social network, despite its weak formalization, is subordinated in the process of establishing trust, knowledge exchange and other forms of behaviour, which are much more influential than administrative orders or managerial decisions (Granovetter, 1973). The functioning of the entities in the environment is an essential field of analysis in the study of the organization and consists of research of business entities' cooperation as well as that of public and social entities. In every case, the canvas of the analysis are the existing social relations, being a base for more formal cooperation systems (Kożuch, 2011).

Researching the practice of analysing networks involves identifying the actors and the relations between the actors. Moreover, it also involves defining and measuring the variables characterizing the network. Very useful in this area is the software for social network analysis (SNA) (Borgatti et al., 2013). Reflective practice in this field refers to the following questions about the cognitive usability of the acquired knowledge:

1. What is the relationship in a social network? (familiarity, knowledge exchange, joint projects, mutual inspiration) and how should the relationships be measured?
2. How should the dynamics of the relationships in the network be considered? (past relations, present relations, future relations, expected relations)
3. How should the network be considered from the perspective of management? (management of an organization in the network, management of the network, utilization of existing relations for management).

Fig. 1. Graphical representation of a social network and activity network

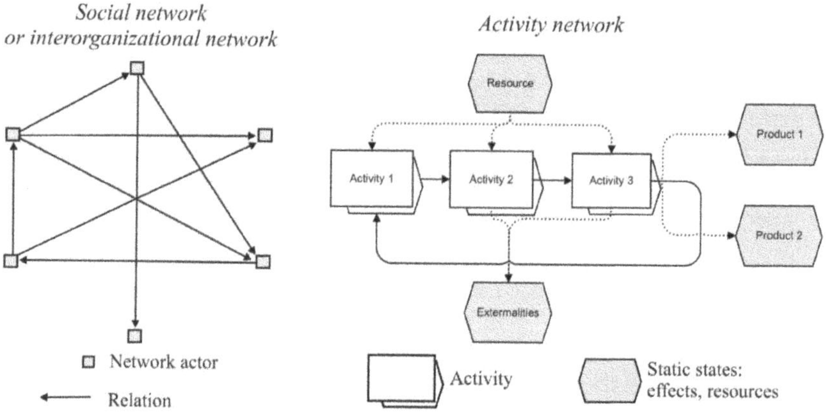

Source: author's model.

In the activity network, the basic element (of the network node) is the activity undertaken by the network's actors, considering the relationship of the interdependence between activities. The questions of the usability and rationality of activities undertaken by people lie in the spotlight of the praxeologists, starting with von Mises, who argued that, for analyses of a category of activity, it is necessary to consider the following aspects: the ends and means, values, human needs. The author considered the exchange as an activity of special value for society, increasing the value of individual activities and therefore the activity in the activity network (von Mises, 1996). Similarly, K. Weick suggested analysing activities in the context of interpersonal and interorganizational relations and as a result of the conscious intentions of people and communities (Weick, 1979).

Activity network theory is the theory of a particular area of substantive actions describing the fundamentals of undertaking informed activities by people, especially in the field of learning, knowledge creation, creativity, but also with any kind of performed work. We can talk about the activity network when for the expected result it is necessary (or the intention exists) to undertake the activities performed by different actors of the network (Engström, 1987). Such phenomena are typical for clusters and networks in creative industries, which are characterized by loose relations and the activity connected with creativity. Collective creativity requires the cooperation of different types of entities creating and transferring generated values as well as organizing the transfer. We are considering creativity at the highest level of analysis — the creativity of individuals in networks connecting organizations, which consists of the aspects of boundary-spanning, brokering, epistemic community, local and global networks, alliances, project-based organizations and other problems (Belussi & Staber, 2012). Creative activities in the networks consist of the creation of a new product by individual members of the network, which has been investigated by P. Gloor in the micro-scale of a project network. The author analysed the way individual members of a collaborative innovation network (COIN) complement a new computer program, by observation of the individual activities of network members (Gloor, 2006). If we visualize Gloor's approach, we will have the activity network presented in Figure 1.

B. Czarniawska presents a network of activities as the structure of interrelated activities which can be linked to the different actors creating the network. The main feature of the social network activity is the domination of individual intentions over the collective intent. However, the author assumes that individual activities are informed by the overall logic of the network (Czarniawska, 2010). R. Miettinen proposes the analysis of the organized innovative activity as a consequence of the network of actors who undertake them, calling such a structure the network of activity systems. Main or grouped activities are the subject of analysis, the graphical presentation of which resembles a production diagram flow (Miettinen, 1999, p. 189). J. Stachowicz argues that the analysis of the actor network and the activity network should be integrated in the social network of activities that consists of five elements for in-depth analysis: the purpose of the network, the set of activities in the network, the set of relations in the network (influence and flows), the function of the location of the network nodes in time and space, and the function defining the level of conscious human activity in setting the network (Stachowicz, 2014). The integrated approach used by the author is useful considering that the relationships in the networks are the result of realized activities. However, they are difficult for analysis because of the high level of complexity of the network.

The concept of activity network can be successfully applied to the analysis of the business process as was presented by N. Viswanadham and S. Kameshwaran. Clothing production realized by a network of suppliers and producers in Asia requires the establishment of dependent activities concerning production, information and management in order to achieve the desired effect — a manufactured assortment of clothing ready on a specified date and to an agreed level of quality. Analysing this process, the authors use the concept of activity network as a canvas for describing the activities of transferring materials and information. This type of business network is a structured activity network being orchestrated by the manufacturer (Viswanadham & Kameshwaran, 2009).

Among the applications of activity networks in management, it is worth highlighting those approaches in which the states (or effects) presenting the organization's performance are linked with the key activities. In such an approach, we are not focused on actors performing activities but on the overall logic of the system being an expression of the organization's strategy. An example of such an approach is presented by F. Rothaermel using the case of capital group Vanguard Inc. According to the author, the value chain creation and activity network are closely related. The strategic activity network (or strategic activity system) of the Vanguard Group consists of interconnected activities and effects. Of interest is the dynamic change of the network of activities in 2011 compared to the situation in 1997. The strategic behaviour of the organization is expressed in the identified network of activities creating a value for the environment. The identification of both positive and negative activities, together with the description of the consequences of these activities, is very valuable to Rothaermel's approach, for example: a lack of relationships with brokers and dealers, a lack of commission for brokers. Very important in this activity network analysis is the clustering of activities – there are only six main activities, while all others are concentrated around them (Rothaermel, 2012).

Rothamel's approach to expressing the organization's network activity is very close to the set of activities representing the business model – an overall concept of business organizations' activities. According to A. Osterwalder's concept of business model, it is comprised of nine elements, among which we can find key activities – repetitive, interdependent activities performed to create and deliver value to the customer (Osterwalder & Pigneur, 2012). R. Casadesus-Masanell and J.E. Ricart give the example of a developed network of activities for chosen airlines, created in accordance with the concept of Rothamel by not only the identification of the dependence of activities and its effects but also by the identification of links between activities and other elements of the business model (clients, cost structure). The authors recommend the establishment of a network of activities before setting up elements of the business model canvas (Casadesus-Masanellandi & Ricart, 2010, p. 198).

Proposal for the methodology for analysing activity networks in networks and clusters in creative industries

The approach to the analysis of the network of activities presented in the literature has been used to develop the author's method of analysing the network of activities in creative environments like creative networks and clusters. The proposed method of exploring the network activity is based on the identification of, and testing, the competence in the network and clusters in order to rationalize the cluster's development strategy. The competence of a certain entity is understood as the ability to perform a certain function in the network, which is valuable for the end-user. In contrast to the competence in administrative science (understood as responsibilities of a particular position in organizational structure), the competences in question are the components of organizational knowledge, while competence (or competence-based) management is the component of organization's knowledge management. The term key competence was introduced by Prahalad and Hamel (1990) for the identification of such resources of an organization which are crucial for the organization's competitive advantage. M. Bratnicki proposed the methodology for diagnozing the organizational competence within an organization and formulating a strategy based on these diagnoses (Bratnicki, 2000). The author of the present article developed the detailed methodology for analysing competence in clusters and networks, which enables the identification of competence gaps, areas of competence duplication as well as formulating the network strategy (Olko, 2014).

The proposed method is one of the fields of research into creative clusters of creative networks. However, it could be implemented for analysing multilateral ventures undertaken by the entities without any established formal relations. In the research project entitled *The models of knowledge management in networks and clusters of creative industries in Poland and the EU countries* the following areas of research were explored:

1. An analysis of the structure of the network using Social Network Analysis (SNA) tools — answering the question what kind of relationships are in the cluster and what are the values of the parameters describing the network/cluster (Borgatti et al., 2013).

2. An analysis of the competence in the network – answering the question as to what are the key competencies of the cluster participants and how they contribute to the cluster's success (Bratnicki, 2000; Prahalad & Hamel, 1990).
3. An analysis of the structure of the network of activities – answering the questions what is the way of creating value in the network, utilizing available resources.

The basic assumption for analysing the activity networks is the utilization of declarations and descriptions of the people involved in the cluster activity, without the interpretation of results before the preparation of the activity network map. According to the assumptions of the grounded theory, the essence of findings is gathered directly from the studied entities – in this case, the actors of the network, using their own language for description the process (Charmaz, 2000). The researcher in the phase of gathering information does not interfere with the content and does not interpret nor generalize information about activities. During the development of the activity network map, the main task of the researcher is to show and verify the relation between activities and states (resources, results).

The detailed research procedure includes the following phases:

1. Preliminary description of activities based on declared competencies and assumptions for the cluster strategy.
2. Interviews with the key actors of the cluster/network to identify key activities.
3. Development of the map of the activity network in the form of activity network diagram with the identification of external states (resources, effects) – Fig. 2.
4. Verification of the map of the activity network in interviews with the main actors of the cluster/network.
5. Analysis and interpreting the results.

Fig. 2. Example map of the network activity developed for the creative cluster

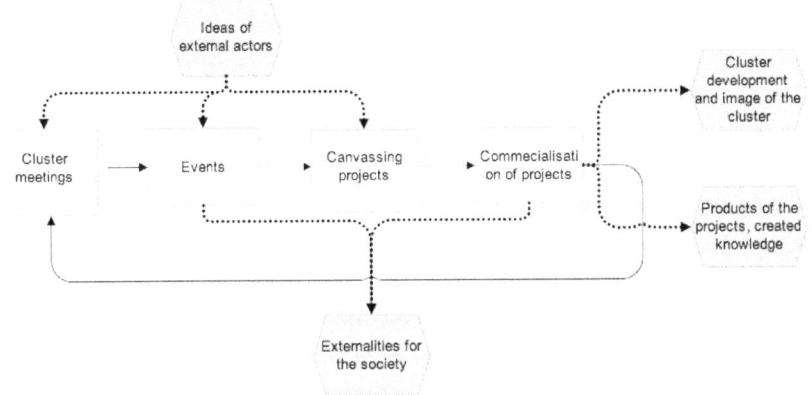

Source: author's own research.

The basic source of information is a semi-structured interview with a representative of the main cluster actors, including a cluster coordinator, which should be performed twice: to identify activities (point 2 of the procedure) and to verify a prepared map of the activity network (point 4 of the research procedure). Figure 2 shows an example map of the activity network prepared during the preparatory research in one of the creative clusters in central Poland.

The map of the network activities can be prepared according to the standards of business process diagrams. The similarities between the activity network analysis and the process approach is not accidental — in the phase of visualizing the network of activities we can use the same standards as in the process management. The map of the activity network in Figure 2 has been created using Microsoft Visio 2007, compatible with the SAP standards for creating an event-driven process chain — EPC. The activity network can be effectively analysed and diagnosed using these standards and tools. However, it is necessary to emphasize the fundamental difference between an activity network and a business process — an activity network cannot be managed like a business process. The global activity of the network of activities results from the conscious and voluntary activities of the network actors. To some extent, it could be the subject of coordination, but it is mainly a self-organizing network focused on the actors' values of interest.

Analysis and interpretation of the results in the form of a map of the activity network (point 5 of the research procedure) should concern the following details:

- the identification of key activities creating added value for the cluster actors on the basis of the evaluation by cluster members (point 4 of the research procedure);
- the identification of the activities not linked to the others (outside the activity network), verification of their usefulness, eventually reduction in not value added activities (NVAA);
- the assessment of the available and needed resources and competences necessary for the efficient realization of cluster activities.

Analyses in this area can be performed using SNA techniques and software: in this case, the identified resources are the nodes of the network (not the actors network), and the relationships represent the links between activities. In the practice of cluster management, an example of activities in the network can be voting – cluster members would vote in this way for the activities which are essential and which of them could be abandoned. But the best way of activity verification is the practical realization of activities identified on the map of activities. Such evaluation requires repetition of research or a continuous monitoring of realized activities in the form of participant observation. It follows the common problem of qualitative research concerning the researcher's reflection on his own subjectivity – the researcher as a human instrument being both a researcher and a respondent (Guba & Lincoln, 2008). The analysis of the activity network like other qualitative methods applied in management science is characterized by many other features, such as the multiplicity of methods and connections with other areas of organizational analysis (Sułkowski, 2012).

Considering the characterized methodology in a reflective way, we can find the essence of the network/cluster analysis in the methods of analysing organizational structure developed by the Aston Group. Operating at the Aston University in Birmingham, a group of researchers led by Derek S. Pugh had proposed and operationalized for organizational structure their own method for the analysis, involving five different dimensions: specialization, standardization, formalization, centralization and configuration. Investigating the network of activities by using the described method is, in fact, analysing the dimensions of specialization and configuration (Pugh et al., 1968).

Conclusions

The methodological approach presented in this paper will be implemented in the research carried out in selected creative clusters in Poland. The conducted preparatory research in two creative clusters made it possible to verify both cognitive and practical usefulness of the method. In one of the explored clusters, analysis or identification of activities have not been conducted at all; in the other one, the analysis of creating an added value using the Osterwalder model has been performed. In both cases, the established map of network activities has been assessed as usable, especially for the presentation of cluster's way of acting. Cognitive usability will be verified after examining a larger number of creative clusters in Poland.

There is one crucial limitation to the implementation of this method, namely – a limited range of activities in creative clusters. Such a state of affairs is evidenced by the results of research carried out in Polish clusters in the period 2008–2010 (Knop et al., 2013). Despite the coordinators' declaration of the high level of cluster's members' activity, their mutual relations were very relatively limited. Also the number of joint projects undertaken in cluster's member was very low compared to the number of ideas for new projects or products. Considering the range of functioning of creative clusters in Europe and in the world, as well as the future of innovation and cluster policy in Poland, we can confidently say that clusters in Poland would be a subject of analysis as active and vivid networks of activities, structuring them using the criteria of value creation in the network. In this perspective, the implementation of the activity network model for the analysis of management processes in creative clusters will provide both cognitive (scientific) and practical values.

Acknowledgements

The article presents selected results of the research project entitled "The models of knowledge management in networks and clusters of creative industries in Poland and the EU countries". The project was financed by National Science Centre on the basis of the decision number DEC-2012/07/B/HS4/03016.

References

Belussi, F., Staber, U. (Eds.) (2012), *Managing the Networks of Creativity*. Routledge.

Borgatti, S.P., Everett, M.G. & Johnson, J.C. (2013). *Analysing Social Networks*. Sage Publications, London.

Bratnicki, M. (2000), *Kompetencje przedsiębiorstwa. Od określenia kompetencji do zbudowania strategii*. Agencja Wydawnicza Placet, Warszawa.

Casadesus-Masanell, R., Ricart, J. E. (2010), From Strategy to Business Models and onto Tactics, *Long Range Planning* 43, 195–215.

de Bruijn, H., ten Heuvelhof, E. (2008), *Management in Networks: On multi-actor decision making*. Routledge.

Engström, Y. (1987), *Learning by expanding. An activity-theoretical approach to developmental research*. Orienta-Konsultit, Helsinki.

Charmaz, K. (2000), *Constructing Grounded Theory: A Practical Guide Through Qualitative Analysis*. Sage Publications, Thousand Oaks, CA.

Czakon, W. (2012). *Sieci w zarządzaniu strategicznym*. Oficyna Wolters kluwer business, Warszawa.

Czarniawska, B. (2010), *Trochę inna teoria organizacji. Organizowanie jako konstrukcja sieci działań*. Wydawnictwo Poltext, Warszawa.

DCMS (2006), *Creative Industries Statistical Estimates Statistical Bulletin*, London, UK: Department of Culture, Media and Sport.

Giddens, A. (1984), *The constitution of society: Outline of the theory of structuration*. Polity Press, Cambridge.

Gloor, P. A. (2006), *Swarm Creativity. Competitive Advantage of through Collaborative Innovation Networks*. Oxford University Press.

Guba, E. G., Lincoln, Y. S. (2009), *Kontrowersje wokół paradygmatów, sprzeczności i wyłaniające się zbieżności*.(In:) N. K. Denzin, Y. S.Lincoln (Eds.), Metody badań jakościowych. The Sage Handbook of Qualitative Research, 3rd Edition, Wydawnictwo Naukowe PWN, Warszawa, pp. 283–289.

Klasik, A. (2008), Budowanie i promowanie kreatywnej aglomeracji miejskiej. *AE Forum 27*, Katowice.

Knop, L., Stachowicz, J., Krannich, M., Olko, S. (2013), *Modele zarządzania klastrami. Wybrane przykłady*. Wydawnictwo Politechniki Śląskiej, Gliwice.

Kożuch, B. (2011), *Nauka o organizacji*. Wydawnictwo CeDeWu, Warszawa.

Luhmann, N. (1995), *Social Systems*. Standford University Press.

Miettinen, R. (1999), The Riddle of Things: Activity Theory and Actor-Natwork Theory as Approaches to Studying Innovation. *Mind, Culture and Activity 6*, pp. 170–195.

Niemczyk, J. (2013), *Strategia: od planu do sieci*. Wydawnictwo Uniwersytetu Ekonomicznego, Wrocław.

Olko, S. (2014), Badanie kompetencji w sieciach i klastrach w przemysłach kreatywnych. *Zeszyty Naukowe Politechniki Śląskiej, Seria Organizacja i Zarządzanie z. 76*, Gliwice.

Osterwalder, A., Pigneur, Y. (2010), *Business Model Generation*, self published, PMBOK A *Guide to the Project Management Body of Knowledge* (2012), Fifth Edition, PMI, USA.

Rothaermel, F. T. (2012), *Strategic Management. Concepts and Cases.* McGraw-Hill.

Stachowicz J. (2011), Globalne sieci przepływu kapitału, wiedzy oraz wartości jako kluczowe wyzwanie w zarządzaniu przedsiębiorstwami, *Zeszyty Naukowe Polskiego Towarzystwa Ekonomicznego 9*, pp. 201–214.

Stachowicz, J. (2014), *Podejście sieciowe (paradygmat sieciowy) w naukach zarządzania; założenia oraz konsekwencje dla praktyki zarządzania.* (In:) J. Stachowicz, M. Nowicka-Skowron, L.A. Voronina, (Eds.), Rozwój organizacji i regionu wyzwaniem dla ekonomii i nauk o zarządzaniu. TNOiK Dom Organizatora, Lublin.

Sułkowski, Ł. (2012), *Epistemologia i metodologia zarządzania*, Polskie Wydawnictwo Ekonomiczne, Warszawa.

Viswanadham, N., Kameshwaran, S. (2009), Orchestrating a Network of Activities in the Value Chain, *5th Annual IEEE Conference on Automation Science and Engineering Bangalore,* India, August 22–25, pp. 504.

von Mises, L. (1996), *Human Action. A Treatise on Economics.*, Fourth Edition, Fox and Wilkies.

Weick, K. (1979), *The Social Psychology of Organizing*, Reading Mass., Addison-Wesley.

Anna Kaczorowska, PhD
*Department of Computer Science, Faculty of Management,
University of Łódź*

Reflections on the effectiveness of the evaluation phase in public projects

Abstract: This chapter characterizes the criteria of time-cost evaluation and their assessment in the following methodologies: Project Cycle Management, Project Management Body of Knowledge and the PRINCE2. The analysis made it possible to formulate recommendations regarding the practical use of the presented methodologies in public administration.

Keywords: effectiveness, efficiency, project management, methodology, evaluation

Introduction

Apart from bearing characteristics typical of other projects, public projects have their own specificity. However, regardless of the type of the project and its background, it is the way of project management that is considered as the key factor influencing the success in all ventures. After all, the success of seemingly similar projects depends on the applied methodologies, techniques, methods and specialistic tools as well as information systems employed in venture management.

Methodologies – the proven standards – are usually a profound source of knowledge about project management. They allow a definition of the benefits to be achieved as well as the products or services by which to achieve them. They should also enable a control of the production process, increase the chances of obtaining the demanded benefits and guarantee gaining and documenting the experience that, in the future, will make it possible to reduce the problems which occurred during the project's execution. However, public administration entities lack general knowledge of the best project management standards, and consequently – effective project management involving a reliable evaluation.

Evaluation is aimed at a survey of project achievements. The point is to make use of the experience gained during its execution in future ventures. The use of knowledge about conducting public projects is limited, because so far no register of implemented venture experiences, to which the interested offices would have an access, has been constructed.

A priority criterion in the selection of the contractors of the greatest projects in the public sector should be the capability of accomplishing such ventures by the company as a whole. However, the Public Procurement Office as the central institution most directly connected with public projects should first determine the project management standards as well as the ways of evaluating the contractors' competences in this field, at best according to project maturity models, such as the Capability Maturity Model Integration (CMMI') or Organizational Project Management Maturity Model (OPM3'), and to supervise their future use.

Significance and relevance of the use of project management methodologies

An increase in the number and complexity of implemented projects creates a demand for management tools enabling an effective accomplishment of these projects. These tools should help the company work out how to repeat its success in future ventures. The project success repeatability is a condition that "has become a premise of every attempt at unifying and standardizing the project management methods within enterprises, organizations, or even branches" (Wyrozębski, 2011a, p. 99). The activities consisting in recognizing, collecting and selecting – in a group of experts – the best practices, well-tried tools and approaches yielded many methodologies of project management. Their multiplicity reflects a certain independence in project management.

According to T. Pszczołowski, methodology means a "methodologically correct set of directives, pointing to ways of action, methods leading to a given purpose, e.g. organizational methodology indicating which methods and techniques are used while performing specific organizational tasks" (1988, p. 119). In turn, M. Trocki postulates that the "methodology of project management is a logical and coherent set of detailed recommendations on how to handle the procedures to be applied for management of the whole cycle of the project, aimed at obtaining the intended result of the project" (2011, p. 17).

The relevance of the development and implementation of project management methodologies is confirmed by both project-establishing organizations and users, i.e. project managers, members of project teams, members of steering committees, and other co-participants of the project process. The main advantage that the organization can derive from the use of methodology is the arrangement and standardization of procedures and the introduction of uniform and identically understood terminology related to project management. Project managers account for the relevance of the development of methodologies by the possibility of using the well-tried practices instead of using in each project the style of management and individual ways of management of particular areas of the project.

H. Kerzner (2003) considers the process repeatability as an important aspect of methodological approach to project management. He states that the project success must be repeatable, achieved by the best management practices, and not just obtained sporadically owing to individual heroism.

The project management methodology determines a set of project execution phases and stages in the project's entire life cycle. However, within each phase the methodology indicates a set of specific results and products of the project as well as a list of activities which will allow to provide them with specific methods, techniques and tools helpful in the project performance.

The phase of evaluation in universal, global project management methodologies enjoying an important status and international recognition, will be the subject of a detailed analysis. I shall address the following textbooks:

1. *European Commission: Project Cycle Management.*
2. *Project Management Institute: Project Management Body of Knowledge* (PMBOK the fifth edition valid from the 1st of January 2013).
3. *Office of Government Commerce: PRINCE2* (version of 2009).

Project evaluation and its types

The term project evaluation should be understood as "(…) the assessment of quality, the extent of a project's execution, as compared to the earlier defined criteria according to appropriate information" (Trocki & Grucza, 2007, p. 225). Evaluation refers to long-term effects, impact and results of the project. According to Nadskakuła, project evaluation is "a complex system of assessment of the project value in view of financial, utility respects, and the accomplishment of the organization's strategy" (2009, p. 9). Evaluation should be conducted in compliance with the rules of utility, feasibility, correctness and accuracy (Trocki, 2012), and its results should assure:

- identification of sources of failures and successes;
- determination of the degree of accomplishment of activities, as compared to the adopted criteria;
- supporting the processes of decision-making;
- increasing the level of effectiveness and efficiency of the use of resources;
- acquiring the knowledge of organization and improvement of project management processes.

The classifications of project evaluation from the perspective of its implementation, subject and function are presented in table 1.

Table 1. Types of evaluation

TYPES OF EVALUATION			
Time of implementation	**Subject of evaluation**	**Function of evaluation**	**Evaluating entity**
Ex-ante Otherwise referred to as the system evaluation. It is conducted before launching the project or programme to find out to what extent the undertaken project will conform with adopted assumptions and whether or not some long-term effects may be achieved.	**Global** Comprises the entire intervention area.	**Formative** In this one the attention is focused on the analysis of processes and their improvement. It supports decision-making processes with the intention to improve the future projects.	**External** The evaluating entity is an external organization. This type of evaluation is performed with a hope for objectivity and high quality of studies for a relatively low price. What arouses anxiety is the evaluating entity's insufficient knowledge of the specificity and background of the tested organization.
Mid-term It is carried out in mid-term of the project or programme implementa-tion, therefore it creates a possibility to modify assumptions which arose during the phase of preparation. It is aimed at investigating the estimated effects of undertaken projects, their results, level of goals achievement, quality of used financial means.	**Horizontal** Otherwise referred to as **thematic** evaluation, because it is focused on a specific thematic area.	**Summing up** It is carried out for the analysis of results of the project as a whole. It supports the decision- making process related to further continuation or withdrawal from the project implementation.	**Internal** One of the organizations directly or indirectly connected with the project is the evaluating entity. The advantages of such evaluation may be: high involvement and factual knowledge. The risks for its results correction are: the lack of objectivity and shortage of people with the required skills.

TYPES OF EVALUATION			
Time of implementation	**Subject of evaluation**	**Function of evaluation**	**Evaluating entity**
Ongoing It is conducted during the project execution, usually after deviations are found in significant parameters of the project.	**Detailed** It comprises a selected issue in a single project.		**Autoevaluation** The people directly engaged in project implementation are the evaluating entity, whereas the project contractors are the main evaluation recipients. It is conducted for assessment of the project quality and its implementation. It does not cover the entire project – only its part.
Final is usually made before the final closure of the project and constitutes a kind of lessons-learned. It is aimed at drawing conclusions for the needs of future projects and use of all types of experiences from the nearly finished venture.			
Ex-post Carried out after the end of the project, whereas its **successive** variant occurs immediately after the closure of the venture. Usually			

TYPES OF EVALUATION			
Time of implementation	Subject of evaluation	Function of evaluation	Evaluating entity
tests long-term effects, but assesses also the level of achievement of the goals determined at the beginning. Its results and recommendations are used at the stage of future projects preparation.			

Source: The author's own study according to (Trocki, 2012, pp. 272–274).

Before the venture is launched, what should be assessed is: firstly – relevance, secondly – feasibility, and thirdly – project *ex-ante* effectiveness, determining the ratio of intended results to useful and planned expenditures incurred to achieve them.

Efficiency, effectiveness, economy and utility of projects

Evaluation of a project may refer both to its results and course. It may be also performed at the following levels of the project management system:

- operational (of the project) – within which the project requirements are compared with achieved results;
- tactical (of the organization) – within which the primary objectives are compared with the results (useful results);
- strategic (of the organization's environment) – at which the general objectives are compared with the main results.

Ex-post evaluation is conducted after the project is completed, i.e. after the results are obtained. The main subject of evaluation is the project's effectiveness which may be also considered at the three mentioned levels. At the project level, "the project's operational effectiveness is determined to find out whether or not, and to what extent, the obtained products are consistent with operational objectives and requirements specified in the project's definition" (Trocki, 2012, p. 279). At the tactical level, in turn, the basic effectiveness of the project is strongly determined to answer the question: are the results – and to what extent – consistent with the basic objectives? The strategic level enables a thoroughgoing evaluation of the venture according to the project's strategic effectiveness criterion (otherwise referred to as the project's impact). The results of this evaluation indicate whether – and to what extent – the main results contribute to achieving the main objectives. The terms "effectiveness" and "quality" of the project are interrelated. If the qualitative requirements contained in the project definition are treated as the project's planned quality, and the project's quality specifies whether – and to what extent – the qualitative requirements were fulfilled, the earned quality is tantamount to the operating effectiveness of the project.

After the project is completed, the project's effectiveness *ex-post* evaluation is made, as determined by the ratio (mostly quotient) of the project's useful results and expenditures incurred on them. If the ratio of the project's useful results and expenditures is calculated as a difference, it is called the project's profitability. Besides, the term "project's economy" is used. The economic project is a venture in which the quotient of useful results to expenditures (costs) is higher than unity. Two cases of the project's economy are singled out: saving and output. The last evaluation is the assessment of the project's utility. This assessment determines whether – and to what extent – the achieved results meet the identified needs of the users and other stakeholders of the project (Trocki & Juchniewicz, 2013).

Evaluation criteria and their assessment in PCM methodology

The phase of evaluation is devoted to the final assessment of the venture's success by the financing institution, which evaluates the project, comparing its planned results with its actual achievements. In the final phase, the comparative data constitute the most important source of information to carry out evaluation related to the strictness of the compliance with the hierarchical structure of objectives in the project's plan. The project's goals are defined according to SMART method (Wysocki, 2013). The criteria used in the EC during evaluation are:

- relevance – the relevance criterion refers to the relevance of the project's objectives for the problems which the project was to solve and the adequacy of these objectives for the physical and political environments in which the project functions;
- preparation of the project and its plan – this involves the validation of the logic and completeness of the project's planning process and the internal logic and completeness of the project's plan;

- efficiency – efficiency is checked by the amount of costs and velocity of management, with the use of which the input and activities were converted into the results achieved in compliance with the required quality level;
- effectiveness – within the effectiveness criterion, the input achieved due to the project's results in relation to achieving the project's objective as well as the impact of the adopted assumptions on the project achievements were evaluated;
- impact – the impact criterion serves to evaluate the effect which the project exerts in a wider environment and its input into more comprehensive sectoral objectives presented briefly in the project's general objectives;
- sustainability – using the sustainability criterion, the EC checks whether or not the services provided within the project exhibit a sustainable nature enabling their further provision for the target group when an external financial assistance is no longer provided.

The third level in the objectives hierarchy refers to the efficiency of activities within the project and is evaluated according to the indicators of effectiveness. The second level in the objectives hierarchy includes the indicators of achieving sustainable benefits for the target group. If the general results of the project do not thoroughly correspond to the plan, further research should be carried out to enable answering the questions:

- Were the poor results caused by the problems which arose during the problems' preliminary analysis, or are they associated with the project's implementation?
- Were the project's institutional and managerial capacities evaluated in a sufficiently precise manner?
- Was the strategy supported accurately?
- Were the financial possibilities to implement the project assessed in a precise manner?

The first level in the hierarchy sums up the project's input into comprehensively conceived sectoral objectives. The assessment of general objectives (Fig. 1) may be conducted as part of thematic or sectoral evaluation. The EC assumes planning formal reports of evaluation at particular stages of the project's life cycle. Such reports usually arise in a mid-term perspective (to survey the progress and propose modifications of the plan for the remaining period of the project's implementation) and after the end of the project (to document the resources, results and progress in the accomplishment of the objective with the aim to acquire knowledge about the project to be used for better planning of future projects). Additionally, the reports are supplemented with an *ad hoc* research – thematic studies (for instance, sectoral projects implemented in one country, institutional projects or types of intervention in one region) enabling the evaluation of many projects, the results of which may be associated with the more comprehensive objectives of national politics.

The Logical Framework Approach (LFA) is the main tool of the PCM methodology. It is also the most important tool during implementation and evaluation processes, providing a basis for preparing a plan of activities, constructing a monitoring system and drawing conclusions from the completed projects. However, the importance of LFA is underestimated because of the lack of obligation to include it into the documentation by those who submit applications for the project's subsidizing. Consequently, there were cases of underpreparation or even a complete failure to prepare the logical framework matrix in projects which obtained the funds (Glińska, Głuszyński, Kowalewska & Szut, 2012; Ministry of Regional Development, 2013).

LFA is a tool which should be estimated and surveyed again along with the advancement of the project and changes in the project's implementation background. Therefore, it needs to be dynamic, assuring not only the first structure of the project subject to scheduling and budgeting, but also a continuous updating of the preliminary plan with the actual data about the time and costs of the implementation of activities.

Monitoring checks whether or not the project proceeds in compliance with the plan. It rests upon making comparisons between the actual status of the project and a set of objectively verifiable indicators (OVI). Determining the objectively verifiable indicators in the LFA, we should calculate the work input and the cost of collecting and analysing the information necessary for verification of each OVI separately. Data collection is a particularly important area in view of the cost of OVI data collection. Therefore, we should reserve appropriate funds for this purpose. If no verification source can be specified for a given indicator, or if it appears to be too expensive or complicated for the collection and analysis, then it should be replaced with a simpler and cheaper one.

Poor preparation and planning of projects was diagnosed (Borowska, 2012, pp. 163–164) using the Project Cycle Management (PCM). Feasibility and sustainability of information projects implemented according to the PCM methodology are significantly determined by the inclusion of the following factors in planning: economic and financial feasibility, appropriate technology, ability and willingness to provide project services after the period of support from the benefactor, socio-cultural issues and gender-related environmental conditions affecting the motivation and participation level as well as environmental protection issues.

Figure. 1. Framework of the hierarchy of objectives in LFA.

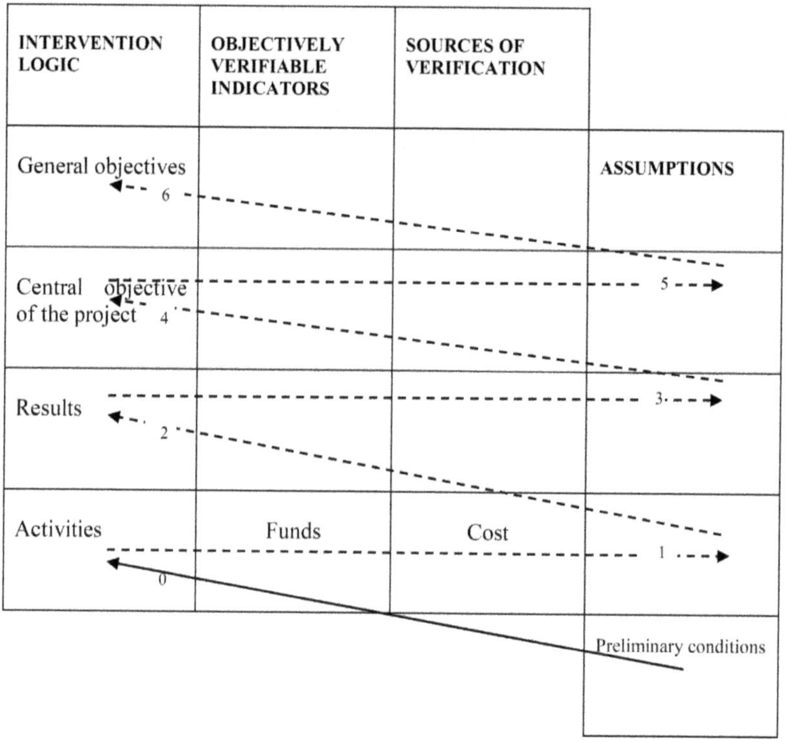

INTERVENTION LOGIC	OBJECTIVELY VERIFIABLE INDICATORS	SOURCES OF VERIFICATION	
General objectives 6			ASSUMPTIONS
Central objective of the project 4			5
Results 2			3
Activities 0	Funds	Cost	1
			Preliminary conditions

Source: Our own study according to (Ministry of Economy and Labour, 2004).

Central objectives of public administration projects are not appropriate for their beneficiaries because they often fail to define the problem in terms of benefits for target groups. We have to make do with an insufficiently detailed definition of the problem when it does not present the real nature of the problem. The level of details defining the problem depends on the scope and type of the project and on the judgment of the moderator and participants of the workshop analysing the problem. In turn, the unavailability of solutions consists in adopting such a definition of the problem which does not present the current negative situation but only describes the unavailability of the demanded situation.

Because of the social nature of most European projects, this methodology provides also solutions which strongly sensitize and integrate the implemented projects with their environment (Grucza, 2011).

However, at the same time it exhibits considerable shortages within the issues directly associated with the project management at the stage of its execution, especially its monitoring. This results from the PCM source based on the concept of management through objectives and the earlier methodology of Zielorientierte Projektplanung (Juchniewicz, 2009). A frequently applied solution is a combination of PCM with PMBOK or PRINCE2 or elements of both of them simultaneously, which makes it possible to provide a relatively complete methodological support for the implemented projects.

Evaluation criteria and their assessment in the PMBOK methodology

Control processes in the PMBOK® methodology are carried out concurrently with executory processes. The prevention against departures from the basic plan, their fast detection, monitoring and the introduction of remedial measures create a possibility of implementing the project in time and without an increase of costs. During the supervision phase of the output products based on which the project is steered, the interim progress reports are submitted (Wyrozębski, 2011b). Elsewhere, I indicate that "closing the project, we should present the project's results in the form of a final report, draw up an independent final review of the project, present the products and give them to the contracting authority to enable their formal acceptance. What is particularly important for implementation of further projects is arrangement and archivization of the project course documentation. The same objective, i.e. drawing conclusions for the future, lies behind the *ex-post* evaluation of the project, which should be carried out by a person (company) independent of contractors and the principal" (Kaczorowska, 2013, p. 223).

The division of the project's life cycle into phases, applied in the PMI methodology, is almost fully reflected in the Microsoft® Project Professional – a specialistic software for management of ventures. The implementation phase (in which the phases of execution and control are combined) consists in an introduction of real data and an analysis of deviations from the basic plan. The achieved progress of work as well as the timeliness and operational costs are controlled not only by the internal indicators of programmes, but also by the indicators of the *Earned Value* method built into the application, which has to be used in the projects subsidized by governmental agencies, ventures accomplished by big organizations and corporations or international programmes financed by the World Bank.

Evaluation criteria and their assessment in the PRINCE2 methodology

Among the methods of complex evaluation of projects, apart from the feasibility studies and business plans of the projects, business substantiation is distinguished. This is a method of assessment of business benefits in the projects concerned, developed for taking decisions by the managers. Business case involves all changes in the business area affected by the project and contains the description of the causes of undertaking the venture, based on estimated costs, benefits, savings and risks. It is used during the project implementation for analysis of the impact of obtained partial results on the expected business benefits.

Business case in the PRINCE2 is considered as a driving force of project implementation. In the PRINCE2 methodology, the business case consists of the following seven elements: premises for undertaking the project, possible variants of implementation, expected benefits of the project, list of the main threats to the project, the project implementation costs, schedule of implementation of a part of the project, and assessment of the cost-effectiveness of the investment i.e. the project. If at a given stage of the project implementation the business substantiation stops being valid, then the project should be discontinued and if the legitimacy cannot be restored the project should be closed completely.

The PRINCE2 methodology has complex and precise procedures of the closure and settlement of the project. It recommends not only the use, in currently implemented projects, of all types of know-how from the closed projects, but also creating a specific register of experiences, to be used in the future in other ventures implemented according to this methodology (Wyrozębski, 2011c).

Maturity of methodologies specialized in project management

The basic part of the project management methodology should be a description of the COMPLETE process of project management. With the aim to classify the project management methodologies, adopted as the criterion of distinguishing the maturity levels was the usefulness of the methodology in the project management practice. This allowed for the following classification of the levels of the project management methodologies (Gasik, 2013):

1. Oriented to functions,
2. Oriented to the following processes: unintegrated, semi-integrated, integrated,
3. Oriented to knowledge: oriented to knowledge units, oriented to attributes, oriented to complete knowledge,
4. Hybrid.

In the earliest period of interest in project management a vast majority of the knowledge referred to the functions which should be performed to implement the project effectively. The practitioners and theoreticians of management first of all took up scheduling and extended the relevant tasks to the time management area. Many techniques and methods – functions effectively supporting such works were worked out to draw up the schedule.

The first to appear was the Gantt bar charts technique, Precedence Diagramming Method (PDM) or Arrow Diagramming Method (ADM), and the Critical Path Method and the last one – the Critical Chain Method. For the project working time management area the following methods of the activity time estimation were developed: parametric, through analogy, bottom-up and top-down. At the stage of their development, the mentioned techniques and methods comprised most of the knowledge about project management, therefore they should be considered to be consistent with the methodology definition. We should also be aware that the methodologies are then equated with a set of techniques and methods, otherwise functions, their development consisting in an improvement of respective functions.

Functions necessary for the project implementation are a very important component of the knowledge about project management. The project manager who uses the methodology containing only descriptions of individual functions (for the area of management of the scope, time, costs and other planes of the project) has many tasks to perform, connected with defining of the process of implementation of the WHOLE project. The same function may occur during various processes, and the processes may be repeated during the project implementation, therefore within an increase in the methodologies maturity the processes definitions were included as their consecutive component. They soon became a basis for the project management methodologies oriented to processes, because they describe a greater part of the management area and in a more complete way support the project managers' work, as compared to methodologies oriented exclusively to functions. Good examples of the methodologies oriented to processes are both PMBOK® and PRINCE2®.

A basis for evaluation of the maturity of a methodology oriented to processes is the maturity of the process description, especially its integration level. Actually, each project has only one process of implementation. To simplify the description of this process, sub-processes and phases consisting of activities are singled out. Sub-processes, phases and activities are interrelated and should be integrated. A non-integrated methodology oriented to processes consists of a set of separate descriptions from various areas of the project management (management of the scope, time and costs).

If in the methodology oriented to processes the integrity management area with its functions and processes is additionally (artificially) introduced, then this type of methodology is called semi-integrated. Its example may be the PMBOK methodology in which the integration management area appeared as the last one in the 1996 edition. The mature methodology containing one coherent process of project implementation from the beginning to the end is referred to as an integrated methodology oriented to processes. An example of such methodology is PRINCE2.

The methodology which apart from the functions and processes involves also knowledge is called the methodology oriented to knowledge. The knowledge of some management areas, e.g. that related to project teams construction, may be used globally, whereas the knowledge of applied price lists is used locally. On the other hand, the knowledge of the risk management area may be used partly globally (this part comprises instability of requirements) and partly locally (the part of knowledge which involves the specificity of a given country or enterprise). Due to their content, the maturity of methodologies oriented to knowledge is higher than the previously presented types of methodologies. The most general way of knowledge presentation is the level of knowledge units. Methodology of such level of maturity indicates only which units of knowledge are necessary for processes implementation. For example while managing the scope of project works we should know the Work Breakdown Structure (WBS method). A directly higher level of knowledge description is the level of attributes. The knowledge units are characterized by the lists of their elementary data, i.e. attributes. Placed at the highest level among the methodologies oriented to knowledge are those including complete knowledge, containing filled in structures from the attributes level.

PMBOK is perceived as a compendium of knowledge, but its authors strongly emphasize that this methodology provides to the project manager a specific set of project management processes, but their final number, form and detailed method of using rest with him.

The only management area common for all projects seems to be the management of the scope and management of time. The other areas may occur but not necessarily occur in particular companies or projects. The need for such a selective approach to project planning, e.g. excluding the cost management area, should be considered as early as at the stage of creation or selection of the project management methodology. The methodologies which enable the choice of a subset of recommendations needed in a given project are referred to as hybrid or component methodologies (Gasik, 2013). A well-constructed hybrid methodology should contain the components of the types of methodologies discussed before. Becoming aware of this fact allows to consider the hybrid methodologies as most mature.

Increasing maturity in project management

The area of increasing the project maturity level contains the ways of the project management improvement. Therefore, to pass from the analysis of the area of increasing the project management maturity to the evaluation of maturity within the improvement of maturity in project management, it was assumed that the evaluation of a methodology will be a maturity criterion for works within the project maturity improvement. At the same time, however, a specialization of methodology in raising the project management maturity was used as a basis for classification into respective levels. This allowed to select the following levels: intuitive, oriented to general management methods, oriented to project management methodologies, and oriented to maturity models.

Some companies implement projects without realizing that the term project exists. Especially the building companies refer to their projects as agreements or contracts and, based on their own experiences and intuition (without methodological fundamentals), they form (functional or matrix) organizational structures and develop improved procedures of activity in particular areas of management. These companies improve their activities through everyday development of better and better solutions based on intuition and experience of appropriate people; therefore this level of advancement in increasing the project management maturity is called an intuitive or zero level (Gasik, 2014).

The general knowledge and techniques used to improve management in an organization are also used for project management. For example, strategic management devised for the general management requirements has been very popular recently in the project management area, where the basic applied approach is the project portfolio management, and the motivation theory is used for the venture's human resources management.

The most frequent example of using the general methods in project management are the quality-related methods, such as ISO 9000 or Six Sigma. ISO 9000 imposes a requirement to write down the knowledge about processes accomplished in the organization, whereas Six Sigma assumes implementation of the continual improvement of the ways of acting, therefore we could assume that implementation of general methods of management raises the project management maturity.

There are methodologies oriented exclusively to project management (PCM, PMBOK`, PRINCE2 or methodologies related to project aggregates – portfolios or programs) and the choice of one of them exempts the organization from a considerable part of specialistic work connected with reaching the knowledge about the project management ways. Taking a decision about implementation of any of the methodologies from this group facilitates achievement of advancement (perfection) in the project management area more than the use of general managerial methodologies (Gasik, 2014). In the article, it is this level of project management maturity – oriented to specialized methodologies – that is the main subject of analysis and evaluation of the effectiveness of evaluation phase in public projects.

The best known maturity models are presently: OPPM3` and CMMI`. The main components of these models are: a list of processes in the project management area, the way of processes classification in view of their maturity, and a list of characteristics of relevant processes in the organization, allowing to qualify the process to one of the maturity levels. Classifications of the processes in the project management maturity models OPM3` and CMMI` are presented in table 2.

Table 2. Processes in models OPM3` and CMMI`

CLASSIFICATION OF PROCESSES		
No	Model OPM3`	Model CMMI`
1	Standardized	Incomplete
2	Measured	Performed
3	Managed	Managed
4	Constantly improved	Defined
5	-	Managed quantitatively
6	-	Optimizing

Source: Our own study according to (Gasik, 2014; Juchniewicz, 2009).

The models OPM3` and CMMI` contain recommendations related to the way of the transition between the maturity levels; the classification of maturity levels may refer to the organization as a whole – step representation in CMMI` or individual processes – continuous representation in CMMI` and OPM3` (Kerzner, 2013).

Conclusions

No methodology will ever become a panacea for all problems faced by project managers, project leaders and project contractors. We should always remember that methodology is only a tool which supports project management, and deciding to choose the existing methodology or being obliged to use a concrete methodology one should aim at its extension by the code of values, organizational behaviour standards and effective information support.

Paradoxically, a characteristic of the integrated methodologies oriented to processes is that no separate functions or processes which integrate other processes are singled out. Methodologies oriented to functions no longer apply to the development of the project management knowledge, whereas the use of hybrid methodologies is planned for the future within public project management in Poland. It seems practically important to differentiate the methodologies oriented to processes from those oriented to knowledge. The implementation of only the project management processes without an appropriate knowledge in this field is an incomplete and ineffective work, because it often fails to solve the problems occurring in project management. The evolution of project management maturity from the intuitive level to the level oriented to maturity models shows a natural sequence of the ways of improving the project management in organizations. A company which wants to improve its project management should not, at the beginning of its activity in this field, implement huge and expensive tools. The implementation of specialized project management methodologies is more effective in those organizations which have already adopted a certain general managerial culture. The implementation of the complete models of project maturity, in turn, is recommended when the organization is known to use one of the specialized methodologies of project management. If it is the organization's aim to reach a complete perfection in project management, then the use of the aim descriptions, contained in CMMI˚ or OPM3˚ (Juchniewicz, 2009), is advisable.

None of the analysed methodologies is ideal and capable of fully meeting the public organizations' needs and expectations. Of the three analysed methodologies, PCM is the only methodology that combines the issues of strategic management with project management. It is a good tool with which the EU institutions may implement – in the form of projects – the earlier developed assumptions and strategic objectives (Trocki & Grucza, 2007). However, a reasonable project manager "condemned" to use the PCM methodology should choose a supplementary set of tools coming from other methodologies or adapt another methodology for his needs.

The analysis of the effectiveness of evaluation in the considered standards shows also that in project management a hybrid of methods should be used. If public administration in Poland does not have its own standard of projecting, then it should know how to choose the tools of appropriate methodologies in compliance with the background of a particular project.

References

Borowska, K. (2012), *Applying the Project Cycle Management methodology in the management of projects co-financed from EU funds*. Obtained from: http://pl.linkedin.com/pub/katarzyna-borowska/85/76/844/de.

Gasik, S. (2013), *Maturity of the project management methodologies*. Obtained from: www.sybena.pl.

Gasik, S. (2014), *Meta model of project management maturity*. Obtained from: www.sybena.pl.

Grucza, B. (2011), *PCM/LOGFRAME methodology: Project Cycle Management*. (In:) M. Trocki (Ed.), *Project Management Methodologies* (ch. 6). Bizarre, Warsaw.

Glińska, E., Głuszyński, J., Kowalewska, A., Szut, J. (2012). *How have the local governments been improving? Evaluation under the name Assessment of competition projects implemented under the sub-measure 5.2.1 Modernization of management in local government administration in Human Capital Operational Programme 2007–2013. The final report for the Ministry of Administration and Digitization*. Obtained from: www.ip.mac.gov.pl.

Juchniewicz, M. (2009), *Maturity of the organizations to project*. Bizarre, Warsaw.

Kaczorowska, A. (2013), *E-services of public administration under project management conditions*. University of Lodz Publishing House, Łódź.

Kerzner, H. (2003), *Project Management. A Systems Approach to Planning, Scheduling, and Controlling*. 8th edition. Publishing House John Wiley & Sons.

Kerzner, H. (2013), *Project Management Metrics, KPIs, and Dashboards. A Guide to Measuring and Monitoring Project Performance*. Second Edition. Publishing Houses International Institute for Learning; Wiley.

Ministry of Economy and Labour (2004), *Manual – Project Cycle Management*. Warsaw.

Ministry of Regional Development, Department of European Social Fund Management (2013). *The letter on the results and recommendations of the ECA control*. Warsaw.

Nadskakuła, O. (2009), *Project Evaluation*. Bizarre, Warsaw.

Pszczołowski, T. (1988), *A short encyclopaedia of praxeology and theory of organization*. Wroclaw: Ossolineum.

Trocki, M. (2011), Methodological fundamentals of project management. (In:) M. Trocki (ed.), *Project management methodologies* (ch. 1). Warsaw: Bizarre.

Trocki, M. (ed.). (2012), *Modern Project management*. Polish Economic Publishing House, Warszawa.

Trocki M., Grucza B. (ed.). (2007), *The European project management*. Polish Economic Publishing House, Warszawa.

Trocki, M., Juchniewicz, M. (ed.). (2013), *Project assessment – concepts and methods*. Warsaw: SGH Publishing House.

Wyrozębski, P. (2011a), Badanie potrzeb i możliwości metodycznego wsparcia dla zarządzania projektami. (In:) M. Trocki (ed.), *Project management methodologies* (ch. 20). Bizarre, Warsaw.

Wyrozębski, P. (2011b), PMI Methodology: Project Management Body of Knowledge. (In:) M. Trocki (ed.), *Project management methodologies* (ch. 4). Bizarre, Warsaw.

Wyrozębski, P. (2011c), PRINCE2 Methodology. In: M. Trocki (ed.), *Project management methodologies* (ch. 5). Bizarre, Warsaw.

Wysocki, R. (2013), *Effective project management. Traditional, Agile, Extreme*. 6th Edition. HELION, Gliwice.

Paweł Romaniuk
Faculty of Law and Administration
University of Warmia and Mazury in Olsztyn

Trends of changes in risk management supported by an internal audit in all bodies of public administration along with their evaluation

Abstract: For some time, the purpose of risk management has been a new challenge for all public finance sector units. Without skilful definition of opportunities and risks related to the implementation of public tasks, it will not be possible to reach the objectives pursued. Proposed regulations, supported by research into public management, provide relevant knowledge and tools for risk management.

Keywords: risk management, internal audit, public administration, regulations, control

Introduction

The evaluation of changes, including risk management, is aimed at setting new standards of behaviour and new trends of research. Furthermore, apart from risk management, we may also talk about the institution of internal audit, which supports the decision making process in the administration. The internal audit allows one to influence the proper implementation of all activities of strategic importance for the organization. The decision-making process in public administration is a procedural and technological feature of the management process, with different economic and psychosocial conditions. The often-evoked rational decision-making model (decision management) is a process which allows for the identification and selection of an appropriate line of action, leading to a solution of a specific problem or an implementation of a new plan. The purpose of this article is to present the essence of risk management supported by relevant management and audit decisions. The author of this article shall depict all the possible trends of changes to accommodate the new perspective of looking at the manner of using the institution of internal audit, which constitutes an added value to change management, decision management and risk management based on professional literature and practical examples. Furthermore, the author shall include in his research the new process approach to the modelling of the system for managing public institutions, corresponding to current trends in management, where the main focus is on the management of processes and audit procedures influencing the legal and organizational situation of public finance sector units.

Risk significance in public finance sector units

We may observe that some risks occur in public finance sector units which make it necessary for the managers to apply appropriate corrective or preventive measures eliminating such risks. Therefore, it is necessary to accurately predict any negative consequences of the activities undertaken within a unit. The current functioning of public institutions results in more and more statutory tasks which are thereby implemented. Undoubtedly, one of the most important processes while realizing the tasks is the proper identification of all events occurring within the unit, whose consequences may negatively affect a given organization. The key step to eliminating such consequences may be accurate planning of the protective system of the unit. The purpose of such a system is usually the prevention of detrimental effects of the aforesaid events as well as the possibility of reducing the risk of the occurrence thereof.

The emerging risk constitutes a significant threat to the ordinary business of public institutions. In such case, the public finance sector unit should take appropriate actions to eliminate the consequences of the aforesaid risk. However, it is also worth remembering that public administration bodies, at all levels of management, should be also fully aware of the positive impact of the risk on the functioning of the units. Such effects of positive risk may increase the involvement of employees in the activities of the unit by offering them appropriate support and adequately preparing risk management procedures (Jajuga, 2007, pp. 12–15).

It may be assumed that the risk constitutes a probability of a certain event that may occur in the unit and result in financial loss or goodwill impairment in a given organization. The risk also means that the unit may not be able to realize certain tasks as defined thereby, which may negatively affect its financial position. The risk entails incertitude, which may be closely related to the events affecting the organizational possibilities of the unit with respect to the realization of the tasks. However, it should be stressed at this point that such incertitude occurs in the events when it is impossible to define and identify all the possible risks. Additionally, the risk is also connected with a specific situation, in which at least one of the decision-making elements remains unknown, yet a probability of its occurrence together with the consequences is known (Michalski, 2010, pp. 34–45). Nowadays, the risk should be understood as a potential variability of events, which is present in all activities performed within the public finance sector units. In such cases, due to the activities undertaken, the public institutions may feel uncertain. However, any exposure to potential risk in the organizations may take place while performing certain activities in the event when, on the one hand, it contributes to a range of benefits, but on the other – to losses.

A lot of attention is also paid to the legislative authority which is responsible for proper governing and administration of the state, and whose decisions (by adopting relevant legal provisions) influence the life of all Polish citizens. Nonetheless, an unavoidable consequence of introducing such new regulations is the fact that they affect many different layers (economic, social, cultural policy). This multilayer action favours the emergence of risks, which is just a matter of time. Therefore, the risk should be identified on the basis of its special characteristics, such as quantification or symmetricalness (Olejniczak, 2009, pp. 16–17). The main focus of the analysis of the essence of risk is always the objective most often adopted by the unit. However, managers of public finance sector units must always realize that the attainment of every goal may bear some risk, which in such case usually means failure to achieve the goal. Every task to be performed by the unit should determine the future goal towards which each organization should strive (Wasilewski, 2010, pp. 25–28). What is more, managers should be fully aware of the fact that the risk may emerge within any area of the activities of the unit. Additionally, an effective prediction method with respect to the risks occurring in the public institutions has become as important as the correct performance of public tasks. Therefore, the application of relevant protective measures, the establishment of safeguards of all the resources and the involvement of the employees in the mission of the organization may, to a larger extent, limit or eliminate any potential risks.

Changes in approach to risk management

Changes related to the functioning of public finance sector units, which treat the possibility of risks in an organization with greater severity, are observed more and more often. Current risk management consists in a special identification and analysis of potential risks. It is especially important to manage such risks together with control activities aimed at assessing any potential threats. In currently existing public institutions, the consequences of risks depend, to a large extent, upon the efficiency and productivity of control procedures, now called procedures of management control, where, pursuant to Article 68 sec. 1 of the Act on Public Finance, management control in public finance sector units constitutes the entirety of activities undertaken in order to ensure the realization of goals and tasks in a legitimate, efficient and timely manner (Act on Public Finance). Furthermore, for over four years now, management control in public institutions has been allowing the realization of seven objectives pre-determined by the legislative authority, i.e.:

1. conformity of business with the provisions of law and by-laws;
2. efficiency and effectiveness of activities;
3. reliability of statements and reports;
4. protection of resources;
5. compliance with and promotion of the principles of ethical conduct;
6. efficiency and effectiveness of the flow of information;
7. risk management (Article 68 sec. 2 of the Act on Public Finance).

The last objective of the control procedure as indicated in the above regulation is risk management, which defines the significance of proper management of all the components of the unit. The risk management process has become a synonym for appropriate goal management by making the right decisions which help to limit any threats related to the possibility of risk occurrence, where rather efficient procedures of management control are aimed at protecting the unit and should also reduce the negative effects of any previously existing risks. The author noticed that the risk management process has already become a concept included in the development strategy of a given public finance sector unit, which also helps to identify any potential events that may have any impact on such a unit. In addition, risk management also provides relevant information and the confidence needed to achieve the assumed goals in an efficient and accurate manner (Mazurek & Knedler, 2010, pp. 32–34).

It is also noticeable that, at least once per year, public finance sector units identify the risks related to their objectives and tasks. Nonetheless, in order to predetermine the areas that may pose a threat to the unit, it would be essential to identify the risks several times per year. This is not an issue of officiousness, but of preventive measures, especially important for larger units that make a number of decisions which usually present higher financial risk. It is recommended, and also more and more frequent, to identify the risks in a cyclical manner in case of material changes in the functioning of public institutions. The cyclical nature of such activities is aimed at eliminating or reducing the possibility of occurrence of such risks. The aforementioned processes are usually defined as preventive and detective control (McNamee, 2004, pp. 124–126).

In the discussed case, we may now distinguish three new approaches to the risk identification process. They are as follows:

1. risk analysis, which examines the impact of the changes and identification of risks that may affect the assets of the organization;

2. environmental analysis, which examines the impact of the changes and risk identification process on the business operation of the unit;
3. risk scenario, where the risks caused by neglecting the control mechanisms and systems are analysed, together with identifying any potential frauds or risks of disaster (Sołtyk, 2009, pp. 62–63).

The risk management policy in public finance sector units also includes a list of areas that are subject to risk identification and assessment. Today's public institutions try to predict the results and consequences of their actions by performing risk analysis. Such a model of prediction allows the organization to prepare appropriate procedures aimed at identifying, analysing, assessing, monitoring, reviewing and making reports about the identified risks for the management of the unit. It should be stressed at this point that the identified risk types must be referred to proper risk management units (Kaczmarek, 2010, pp. 56–76).

Nowadays, the purpose of risk analysis in today's public institutions is to show the significance of the risk and identify the reasons for its occurrence. Undoubtedly, one of the most important issues is to learn about the reasons behind the occurrence of the risk, as well as to apply appropriate corrective measures, which is the duty of the managers of a given unit, where such measures usually concern the reasons for the occurrence of the risk and not its effects. In the discussed risk management process, global assessment of the risk level affecting proper functioning of the entire organization becomes necessary. Therefore, each year, the management of the public finance sector unit defines the scope of potential risks. As part of risk analysis, the unit's management determines the causes and consequences of such risks, using the so-called categorization of existing risks. In such case, we may usually distinguish various types of risk categories (e.g. strategic, operational, decision, financial, HR, internal, external, corruption, over-valuation, residual).

Undoubtedly, another crucial issue accompanying changes in risk management is to define the so-called risk appetite. The risk appetite is used to determine the risk level, which the unit's manager is usually willing to accept. At times, we also talk about the so-called acceptable risk determined according to the level of materiality, which is required to monitor the risk in the event when the organization already has its control mechanisms. The acceptable risk is usually a negligible risk (low impact). The discussed level of acceptable risk depends on many factors, usually external, which derive from both the nearest surrounding and further areas, which remain unaffected. Additionally, some internal factors may also arise, depending on the decisions and activities taken by the unit. It is possible to determine and apply the methods for measuring the risk appetite by the organization, which is fully aware of the existing risks and is mature and able to identify, assess, measure and control any risk. Such public institutions that use their HR and decision-making potential and are able to prepare and apply a risk management system appropriate to the specificity of a given organization are growing in number.

Internal audit and risk management

The main tool used for analysing the efficiency of managing public finance sector units is an internal audit, supported by the previously discussed procedures of management control and risk management. In case of the entire public sector, we may talk about the responsibility of the unit's manager for financial management. The unit's management is fully responsible for the efficiency of management. Therefore, the managers of the organization must know how to skilfully implement the adopted risk management strategy. Additionally, such a strategy must be subject to regular evaluation performed by an independent internal auditor operating within the given unit.

In compliance with the instruction contained in the Act on Public Finance, the internal audit activity should be independent and objective. The most important purpose of an internal audit is to support the minister responsible for the department or manager of the unit in achieving their goals and realizing tasks by way of regular evaluation of the procedures of management control and other advisory activities. This kind of evaluation should be mostly about adequacy and efficiency of management control in the government administration department or local government unit (Article 272 sec. 1 and sec. 2 of the Act on Public Finance).

It must be stressed here that an internal audit is by no means responsible for the risk management process. It means that the main purpose of the internal audit in the field of risk management is proper risk identification and analysis. The aforesaid risk analysis includes, in particular, the identification of the area of activity and risk assessment within all identified areas of activity of the unit, which are called the risk areas. Therefore, while analysing the risk, each manager of the internal audit department or auditor of the service provider is obliged to consider the scope of responsibility of the unit's manager for proper execution of management control (§ 5 sec. 3 of the regulation on the performance and documentation of internal audit activities).

The purpose of an internal audit is mainly to provide the unit's manager with information on the progress in all tasks within the organization. The internal audit process is also aimed at providing the unit's manager with hints and advice so as to increase risk management efficiency. When discussing the role of an internal audit in the risk management process, it is impossible to omit the normative act, which is of great importance for the internal audit department, i.e. the International Professional Practices Framework of the Institute of Internal Auditors, being an appendix to the communiqué of the Minister of Finance issued on 17 June 2013 with respect to internal audit standards for public finance sector units. On the basis of the provisions of the aforesaid communiqué, it is evident that the internal audit must objectively evaluate the effectiveness of the unit's management and contribute to the improvement of risk management procedures. By taking proper actions, the internal auditor considers the risk management processes as correct and efficient, provided that, based on the auditor's evaluation (*Internal audit standards…*, p. 15),

- the adopted goals of the organization support the unit's mission and are in line therewith;
- serious risks have been properly identified and assessed;
- the method for dealing with the risk, which is in accordance with the organization's appetite for a given risk, has been well chosen.

The internal audit process in the field of risk management is aimed at evaluating the governance set-up of the unit, business operations and IT systems, which are most often exposed to the risk of failing to achieve the strategic objectives of the whole unit, reliability and credibility of financial information, efficiency and productivity of the business operations and all programmes, protection of the unit's assets and conformity with the provisions of law, procedures and agreements (*Internal audit standards…*, pp. 14–15).

The activities of the internal audit department in the risk management process also include professional support of the discussed process. This is done in the form of an independent advisory concerning the functioning of the control mechanisms adopted in the unit. The role of the auditor is to check whether the risk management strategy has been prepared and implemented in the unit. The auditor shall also determine whether appropriate employees of the unit have been selected to implement the risk management strategy. When giving his/her opinion about the efficiency of the implemented risk management process and potential recommendations with respect to changes in risk management systems, the auditor must comply with the basic rules included in the independence and objectivity standards. Such standards stipulate that while identifying and assessing the risks, the audit activities should be fully independent, and the auditors who give their opinions about the efficiency of the control mechanisms should also remain independent and objective (*Internal audit standards…*, p. 6).

The internal audit process is also aimed at assessing the risk of fraud and managing the risk in this area. The internal auditors must properly assess the risk of fraud. They do not have to detect such irregularities, but they can. A quick reaction to the aforesaid activities of the internal auditors may be even essential for the unit's managers in order to take appropriate steps as well as remedial and protective decisions. The auditors must use their knowledge of risks obtained while performing advisory tasks to evaluate risk management processes in the organization (Gut, 2006, pp. 88–92).

In the event when the internal auditors help the unit's managers to create and streamline the risk management processes, they must remember not to take over any responsibilities of the management personnel or actually start managing the risks. The basic role of the internal audit process is to provide managers with an independent opinion on the efficiency of internal audit mechanisms concerning risk identification and assessment. However, all the decisions related to the functioning of the organization and its reactions to any risk must be handled by the unit's managers. At the same time, an internal audit may also be included in the risk management process in case of, among other things, the early warning system, which facilitates and channels the risk management process or the activities of the auditors acting as experts in specific fields of the functioning of the organization (Wolska-Hertman, 2005, p. 15).

Conclusions

The risk management policy in public finance sector units defines the risk management methodology. It also distinguishes specific principles that become a useful tool for managing such a process. It should be also clearly stated that this kind of policy lays foundations for the risk management system, which is being implemented and already exists in all public institutions. The key condition for implementing the risk management policy is to define clear tasks in accordance with the mission of the organization, as well as to distinguish indexes used for measuring the targeted goals and tasks of the unit. The risk management policy also means the establishment of the threshold of acceptable risk, where it is now necessary to conduct an ongoing and regular monitoring of the progress in goals and tasks, including the application of management control procedures. It is also worth remembering that the unit's manager is always responsible for implementing the risk management policy. The manager is responsible for shaping and implementing the risk management policy, controlling and monitoring the efficiency of the risk management process, establishing the level of acceptable risk and making any and all decisions related to the manner of reacting to different types of risks.

We may observe that a new model for protecting public funds, i.e. the internal audit process, has been successfully implemented in the public finance sector. Following the experience of the English-speaking countries, which have extensive experience in the field of internal audit, the application of this procedure in Poland has been mostly directed at providing advice and identifying the areas of business activities where the risk of irregularities is considered to be the highest. The managers of public finance sector units start to feel more and more confident that the management systems work correctly, and any irregularities that may occur are eliminated at the outset. The employees working in the unit, who are responsible for implementing the risk management procedures defined in the risk management policy, play a major role in the entire process. This is done by correctly identifying the risks and owners of the identified risks, as well as applying appropriate corrective measures. It must also be remembered that the above mentioned methods of risk assessment shall never fully guarantee that the risk measurement is reliable. However, they provide some knowledge of the prioritization of risks occurring within the unit, which are later disclosed in an annual audit plan. Undoubtedly, the internal audit process, thanks to proper risk identification and analysis, has become an important tool allowing one to better realize the financial tasks from EU funds, already implemented in the new perspective for 2014–2020, and hence becoming the key instrument supporting the risk management process in public finance sector units.

References

Wasilewski, W. (2010), Zarządzanie ryzykiem w zarządzaniu kryzysowym. (In:) P. Antonowicz (Ed.), *Innowacyjne strategie kreowania przewagi konkurencyjnej przedsiębiorstw* (25–28), publ. Fundacja Rozwoju Uniwersytetu Gdańskiego, Sopot.

Gut, P. (2006), *Kreatywna księgowość a fałszowanie sprawozdań finansowych*. C.H. Beck, Warszawa.

Jajuga, K. (2007), *Koncepcja ryzyka i proces zarządzania ryzykiem – wprowadzenie w zarządzanie ryzykiem*. Warsaw: publ. Wydawnictwo Naukowe PWN.

Kaczmarek, T. (2010), *Zarządzanie ryzykiem. Ujęcie interdyscyplinarne*. Wydawnictwo Dyffin, Warszawa.

Mazurek, A., Knedler, K. (2010), *Kontrola zarządcza – ujęcie praktyczne*. Wydawnictwo Handicap, Warszawa.

McNamee, D. (2004), *Oszacowanie ryzyka w audycie wewnętrznym i zarządzaniu*. Fundacja Rozwoju Rachunkowości w Polsce, Warszawa.

Michalski, G. (2010), *Strategiczne Zarządzanie Płynnością Finansową w Przedsiębiorstwie*. Centrum Doradztwa i Wydawnictw CeDeWu, Warszawa.

Olejniczak, A. (2009), Zarządzanie ryzykiem w jednostkach sektora finansów publicznych – podstawy prawne, standardy i sposób efektywnego przeprowadzenia. *Skarbnik i Finanse Publiczne*, (5), 16–17.

Sołtyk, P. (2009), Zarządzanie ryzykiem jako przedmiot oceny audytu wewnętrznego. *Finanse komunale*, (3), 62–63.

Wolska-Hertman, M. (2005), Rola audytora w zarządzaniu ryzykiem. *Gazeta Prawna*, (6), 15.

Act of 27 August 2009 on Public Finance, Journal of Laws of 2013, item 885, as amended.

Regulation of the Minister of Finance of 1 February 2010 on the performance and documentation of internal audit activities, Journal of Laws of 2010, No. 21, item 108.

Communiqué No. 6 of the Minister of Finance of 6 December 2012 on detailed guidelines for the public finance sector concerning the planning and management of risk, Government Gazette of Minister of Finance 2012, No. 29, item 56.

Communiqué No. 2 of the Minister of Finance of 17 June 2013 on the internal audit standards for the public finance sector units, Journal of Laws of 2013, item 15.

Katarzyna Szara, PhD
Department of Economics and Management
Faculty of Economics
University of Rzeszów

Creative capital and capabilities of its measurement within an organization

Abstract: This chapter discusses the theory of creative class and develops a theoretical framework of creative capital in relation to the components of intellectual capital of the organization. As a result, the study provides a conceptualization of organizations' capabilities related to building their internal creative capital.

Keywords: creativity, capital, effects, assessment, methods

Introduction

In general, management is about achieving expected results. However, the results of creative process are difficult to measure, especially because it is often considered reflectively. The challenge pertains to comparing the rather intangible results of the process with the very tangible inputs it requires. The effect of creativity is an idea, manifested later through innovations.

Creativity is related with human needs and is part of the development of an individual. Therefore, the human being is the most important resource of every organization. Within an organization, the creative process requires employing people with particular skills and competences. It also requires favourable conditions and environment. It is important to build creativity from the top to the lowest level within an organization. Therefore, a stimulation of passion and action is a very important factor increasing the level of organizational creativity.

The aim of the chapter is to depict the capabilities of building creative capital within an organization, and the elements of its assessment based on certain evaluation criteria. The attempt was made to build a theoretical framework of creative capital within the organization, which may support its assessment referring to the effectiveness, efficiency, durability and reliability of actions.

Creativity within an organization

Creativity pertains to our everyday life, it is required from many workers, students, etc. In an organizational context, it is often identified with the process of discovery and support of outstanding talents, creating conditions shaping the creativity of people, accepting the risk of applying new solutions to particular problems. Being creative means the ability of imaging that which has not existed yet as well as seeking present solutions and new forms of work. Being innovative means introducing envisioned changes within society and economy (Kozioł & Zaborek, 2012, p. 8).

Three groups of components constitute creativity (Nęcka, 2001):

1. Features of the creation – i.e. reliability, originality, peculiarity, necessity and aesthetic values;
2. Psychological reaction of the recipient – effective astonishment, preliminary distrust, effect of the secondary assessment, the psychological effect, complimentary effect;
3. Features of the thinking process – mobility, synthesis, active relation to creating, breaking the mental block and action.

This definition, to a large extent, extends our point of view on creativity, focusing on many aspects neglected by others (Lipka & Waszczak, 2012, pp. 63–64). Creativity relies on combining knowledges of various fields to create new original ideas. Creativity is about constantly discovering new ways, stating challenges against previous manners of thinking and coping with conflicts which the actions lead to. Creativity relies on discovering new meanings in various aspects of living and experiencing, and their unconventional combinations. The notion of creativity depends on the context, in which the new idea, product or the manner of conduct is offered (West, 2000, p. 12).

Creativity is connected with great subjectivity of assessment. Usually it is another person, group, organization, or society that decides whether someone is creative or his or her work is perceived as such. It is worth remembering that the measurement of the creativity level of an individual does not have to be reflected by specific material products. Creativity is mostly the potential of a human, which makes him or her able to create something new or to use something in a new way, and to produce values for him- or herself and for others (Bubrowiecki, 2009, p. 10).

The definitions of creativity are multidimensional due to their presentation in various fields of science. Usually, they pertain to the following areas (see Wallas, 1926, pp. 16–22; Wertheimer, 1968, p. 9):

- feature, which is proper for each human regardless of age;
- group of values;
- thinking process;
- attitude related to thinking out of the box and allowing to create something new;
- capability resulting from the law of change.

Regarding the broader approach to creativity, it may be defined as a process of gaining and conscious or unconscious exchange of acquired knowledge, life experience, competences used in an unconventional manner in a creative process. The novelty of this definition pertains to embracing a depreciation of creativity, meaning that the process may undergo ageing, a loss of the update. On the ground of economics, the process of changes and renewal of capital is important from the viewpoint of its wear and tear. Material resources wear out from the technical, economical or ecological viewpoint. It may be anticipated in case of creativity, too. If it is an ability of a process character, it undergoes a process of changes.

In theory, the concept of a creative organization is based on the model integrating particular kinds of intelligence and creative awareness appearing within processes, and defined for the employee, team and the entirety of the company (see Brzeziński, 2009, pp. 10, 32–33). According to Brzeziński, such a combination constitutes the concept of creative capital of an organization and makes it possible to establish creative capabilities on particular levels of management. The metaphor of a combined mind of an organization reflects the model based on the combination of the so-called morphic field and the psycho-social attitude of a collective mind. (Brzeziński, 2009, p. 11). It includes certain elements presented in Table 1.

Table 1. Components of creative organization

Individual dimension	Collective dimension
Need for extending qualifications Flexibility of thinking and acting Intellectual abilities and broad cognitive horizons Emotional values Learning creativity Persistent aiming at success Independence of other's opinions and faith in oneself Sense of freedom in thinking and acting Courage to undertake risky undertakings and ideas Skills of holistic thinking Eagerness to break existing paradigms Openness to changes	Diversity of competences and combining them for common good Free acknowledgment of collectively worked out principles based on positive bonds Equal status of partners Assigning transient forms in the implementation of the tasks, project by organizational unions between partners Reduction of hierarchy to the minimum Free information flow Responsibility and self-control in creative teams Diversity depending on the level of maturity of the teams Motivating atmosphere Orientation to environment

Source: Brzeziński, 2009, p. 67.

A creative organization is an organization in which all kinds of intelligence and creative awareness play a dominant role. In the concept of creative organization, the phenomenon of transferring and combining individual processes of the mind into collective processes of a higher rank, influences the information field created within an organization, consisting of creative teams situated on particular levels of management. Such an organization is capable of providing new solutions to various problems, developing unique concepts, original methods, technologies, machinery, products or services. These all functional areas, including management of human capital and marketing, are oriented on creativity (Brzeziński, 2009, pp. 79–82).

Due to the fact that creativity is encapsulated in human capital, and the human capital is included in the non-material assets of enterprises, the economy of creativity refers to the economy of non-material assets and new economy. At the same time, creativity (at least in the so-called behavioural attitude and theory of generativity) is mostly interpreted as an untypical reaction or behaviour. It supports the argument to include creativity economics into behavioural economics. Creativity economics is included not only in economic sciences but also in so-called creatology, along with (aforementioned and following) creatological sub-disciplines. Research on such disciplines within the framework of creativity economics, allows to identify and apply in practice the most efficient and effective ways of increasing creativity of human capital and the absorption of creative goods offered (Lipka & Waszczak, 2012, pp. 24–28).

Creative capital is derivative of human capital. This notion embraces the concept of creative industries, the creative class and the development of cities, among others.

Development of creative capital within an organization in the area of talent, technology, tolerance

In this chapter, creativity is understood and perceived through the lens of creative capital. The conditions indispensable for attracting and retaining human capital of high quality encompass the connection between the most brilliant employees, atmosphere of tolerance and openness as well as the technological level according to the 3T concept (talent, technology and tolerance). The characteristics of the conditions is connected with the creative class, the notion proposed by Richard Florida. The author confirmed that this class gets its identity from the role of creativity suppliers. Assuming that creativity is the driving force of the economic growth, this class became dominant within a society. Despite the fact that Florida differentiates that creative class has a proper force, talent and number, its members are the entire society "having an opportunity of transferring internal seeking their own spirit into real energy leading to renewal and transformation" (Florida, 2010, pp. 21, 23).

Thus, there is a statement that creative economy is the dynamic, stormy and fascinating system, providing their divisions and causing stress. This assumes a new life style preferring individualism, self-determination, acceptance of diversity and desire of rich, multidimensional experiences. Encouraged by the ethos of creativity, creative people combine work with life style so as to create their own identity. It means that people define their identity by means of many numerous creative actions. For many, combining social roles and many interests is the confirmation of exclusive creativity. It also appears in organizations.

The consequence of establishing creative economy was a differentiation of creative groups or classes. This is, according to Florida, an economic class, as the economic function influences social and cultural choices and life style of their members. It is created by people who increase an economic value by their creativity. The creative class includes knowledge workers, symbol analysts, professionals and technical workers, though, their role is played in economy. In the definition by Florida, the emphasis is put on the manner of organizing people into social groups, creating the common identity based on the economic function. The economic function defines their social and cultural preferences, consumer habits and social identity (Florida, 2010, pp. 22, 35).

This is not the division into classes from the traditional ownership viewpoint. The ownership of this class has an intangible character, derived from creative abilities. The creative class is composed of super-creative core and professional creators. Super-creative core is created by scientists, stage artists, actors, designers and architects, poets and novel writers as well as representatives of opinion makers of the contemporary society – authors of fact literature, issuers, figures of culture world, analysts, think-tank analysts, engineers, film architects. The work of those people is connected with solutions but also seeking problems. In turn, professional creators work in the fields requiring advanced knowledge, including doctors, lawyers, managers and their profession requires testing and mastering new techniques, applying new manners of treatment (Florida, 2010, pp. 83, 84).

The largest groups of a creative class are located in cities. It is connected with the conditions which are created for development of representatives of particular professions. The base for such development is depicted by three factors – talent, technology and tolerance (3T).

Different researchers of the creative class noticed that in the narrow understanding it constitutes creative capital. Undeniably, every human is creative to some extent. If one performs work in his or her profession, it also constitutes a kind of creative capital. The difference pertains only the scale of its use by a particular profession. If every person may be creative, then by using this potential creative capital is created within a creative organization. However, this capital is influenced by various external factors, depicted by the following indices (Florida 2010, pp. 83, 343–344, see also Szara, 2014, pp. 86–87, Klincewicz 2012, pp. 86–87):

- coefficient (index) of high technologies is based on two variables – percentage share of the production sector of high technologies from a given region in total production of the entire sector of high technologies in the country and percentage share in the regional production coming from high tech industry in the entire income of a given region (country);
- coefficient of innovativeness is the number of patents in a given region per given year in comparison to the number of dwellers;
- coefficient of gays defines the percentage of homosexual pairs on a given territory in comparison to the entire population in the country;
- coefficient of boheme defines the number of artists in a given territory in comparison to a national average, to the population in the entire country;

- coefficient of talent (defined as the coefficient of human capital) presents the percentage of persons with a bachelor's degree and higher within the entire population of a given region;
- coefficient of the melting point measures the percentage of persons of a different nationality born in a given region;
- coefficient of diversity is the aggregation of the gays, boheme and the melting point;
- coefficient of creativity is the aggregated coefficient of innovation, high technology, number of gays and measurements of the creative class (number of persons performing creative professions to the overall number of the employed).

The studies on the creative class use modified coefficients, which is a result of the accessibility of data. The analysis of the creative class and capital concerns countries, regions and cities. It may also refer to the creative industry, enterprises and organizations. In this manner, creative capital will be analysed in the next parts of this chapter.

The assessment of the conditions predestining a given location to the development of the creative class, based on the number of the representatives of the professional class and indices, does not cause any doubts due to the comparability of the available data. Thus, the analysis of creative capital was proposed, starting from human resources. It has been based on the model of creative capital. The model allows an indication of further directions of studies and assessments (Figure 1). Treating creative capital as the resources of human capital is connected with perceiving a human within an organization. On the basis of the sciences on human capital, management is treated as an element of intellectual capital. Thus, we encountercreativity not only in the very essence of human capital but also in intellectual capital. Creative capital is part of intellectual capital, which is part of human capital.

Creative capital is activated within the framework of self-awareness or relations with other people. Developing our talent, we develop ourselves. We develop ourselves by creating, experimenting, growing, risking, breaking rules, and making mistakes.

Figure 2. Model of creative capital in an organization

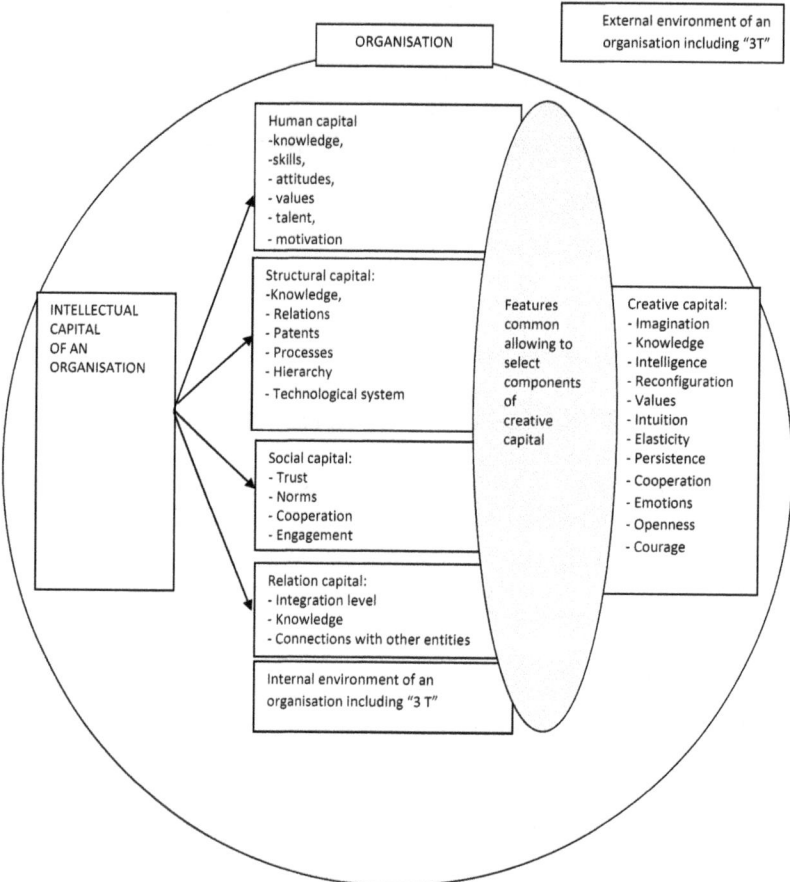

Source: Own study.

The elements of creative capital may be identified among all components of intellectual capital. People, who are attributed with creative capital, should be assigned the features of intellectual capability supported with the courage of action and emotions, which allow one to be open to new thoughts. Moreover, it must be emphasized that creative capital is influenced by many factors, both inside and outside the organization. On the basis of different studies (see Boschma, 2007; Clifton & Cooke, 2007; Florida, Mellender & Oism, 2008), it may be stated that talent, tolerance and technology are those factors that influence its development.

Conceptualizing the dimensions of creative capital evaluation

Evaluation, due to its diverse character, is understood as an assessment. In this case, it is an assessment of creative capital. Such an assessment is of analytical and systematic character. It requires obeying the research rigour typical for social sciences (see Bienias, 2012). Within this notion, three aspects of the investment projects evaluation are particularly applicable in terms of assessing creative capital. Firstly, the analytical and systematic character of a process is emphasized. It means the application of a scientific attitude based on data, and following the research rigour of social sciences, regardless of the quantitive and qualitative orientation of the study or mixed research methods. Secondly, it assumes evaluation of the quality of a given intervention (merit-quality) and also its value in financial and economic categories (worth-value). Thirdly, the evaluation should include the assessment of the processes and actions of the programme/project and their effects (Olejniczak, 2007, p. 16).

Evaluation allows valuing and indicating the directions of solutions on the basis of a set of criteria. They include (Ewaluacja, 2004, p. 3):

- Effectiveness – allows to determine whether the assumed targets have been achieved at the level of products, results or influence;
- Efficiency – relies on comparing engaged resources with achievements at the level of products, results or influence of the programme;
- Relevance – allows to assess the reliability of programme aims with the needs of the sector or region;
- Utility – relies on comparing the needs of the sector or region with the achievements of a given programme;
- Sustainability – relies on defining the durability of the programme results after finishing its financing.

Those criteria are proposed to be used for assessing creative capital (Figure 3).

Figure 3. Proposal of assessing creative capital

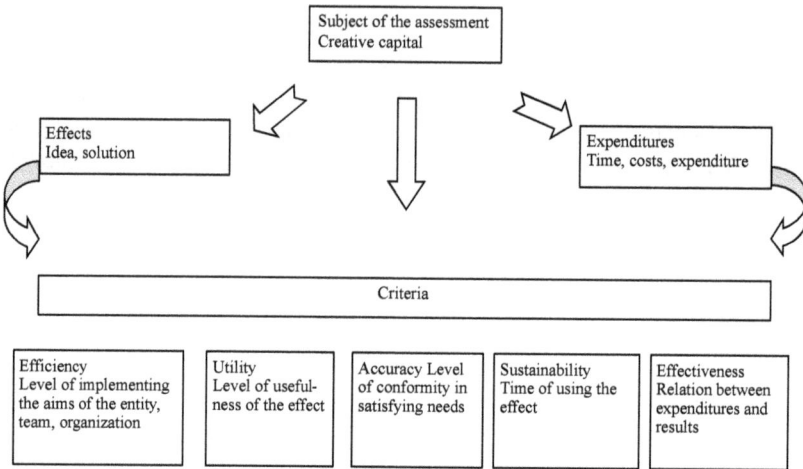

Source: Own elaboration based on literature review.

The presented classification is based on the division according to the two most elementary aspects – expenditures and effects. A human being, developing his or her creative potential, goes beyond the scheme. Such actions are connected with various expenditures: pecuniary, material, time, energy, work. The assessment of such inputs may be performed in the qualitative and quantitative manner. It would be hard to measure the effects of creativity. The effect may rely on an idea transformed into a given form which is innovation. The criterion of utility will define the level of usability for satisfying the needs of an entity, team or organization. Relevance may be assessed by the level of meeting the needs, satisfied by implementing an idea or solving a problem. Sustainability is reflected by the durability of the effect's usefulness. Efficiency is measured by the relation between expenditures and results.

Each of the criteria has its own unique properties. It must be emphasized that the selection of the assessment criteria depends upon the problem which is faced by the evaluator within the research process. It may depend on the research time and scope, or may refer to the assumed paradigm. The outlined evaluation framework makes it possible to provide the methods, tools and means for further assessment of creative capital within an organization.

Conclusions

Being creative means the ability of envisioning something that has not existed yet as well as seeking new solutions and new forms of work. It is not an elite feature, but it can be achieved by every person. The concept of creative capital emerged in literature as the answer to using the distinctly human feature of creativity. This is the proper resource for people, who cooperate and function within new conditions. The development of creative capital pertains to the potential of an organization. In a narrow approach, it is interpreted as the creative class, developing in the favourable conditions. Talent, technology and tolerance are the key determinants of this kind of capital.

Creativity appears to be an element indispensable for any creative functioning. By seeking new unconventional solutions, an individual develops his or her creativity, but if he or she operates within an organization, the creativity of other members and the entire organization may also develop.

Creative capital, due to its intangible nature, is difficult to measure. The outlined framework of creative capital, seen as arising from human capital, differentiates the common features which constitute it. The application of the evaluation criteria derived from investment projects to creative capital emerges as a noteworthy strategy of dealing with at least some measuring challenges.

References

Boschma, R.A., Fritsch, M. (2007), *Creative Class and Regional Growth in Europe – Empirical Evidence from Eight European Countries*. Jena Economic Research Papers.

Brzeziński, M. (2009), *Organizacja kreatywna*. Wydawnictwo Naukowe PWN, Warszawa.

Bubrowiecki, A. (2009), *Sekrety kreatywnego myślenia: jak rozbudzić swoją kreatywność i tworzyć genialne pomysły?*Internetowe Wydawnictwo "Złote Myśli", Gliwice.

Clifton, N., Cooke, P. (2007), *The 'Creative Class' in the UK: an initial analysis*: Regional Industrial Research Report 46, Centre for Advanced Studies, Cardiff University.

Lipka, A., Waszczak, S. (Eds.) (2012), *Ekonomia kreatywności. Jakość kapitału ludzkiego jako stymulator wzrostu społeczno – gospodarczego*. ZN UE w Katowicach, Katowice.

Ewaluacja w PARP. Wytyczne do systematycznej oceny programów realizowanych przez PARP, (2004), PARP, Warszawa.

Bienias, S. (Ed.) (2008), *Ewaluacja. Poradnik dla pracowników administracji publicznej* (2012), MRR, Warszawa.

Florida, R., Mellander, Ch., Qian, H.(2008), *Creative China?* The University, Human Capital and the Creative Class in Chinese Regional Development, Dostępny w Internecie http://www.creativeclass.com/article_library/category.php?catId=53.

Florida, R. (2010), *Narodziny klasy kreatywnej*, NCK, Warszawa.

Klincewicz, K. (Ed.) (2012), *Klasa kreatywna w Polsce. Technologia, talent i tolerancja jako źródła rozwoju regionalnego*, UW, Warszawa.

Kozioł, G., Zaborek, R. (2012), *Przedsiębiorczość drogą do nauki kreatywności i pracy zespołowej*. Urząd Marszałkowski Województwa Wielkopolskiego; Poznański Akademicki Inkubator Przedsiębiorczości. Poznań. Dostępny w Internecie:iw.org.pl/kreatywnyuczen/media/.../Podręcznik_użytkownika_modelu.pdf (online, access: 2014-04-08).

Nęcka, E. (2001), *Psychologia twórczości*. Gdańskie Wydawnictwo Psychologiczne, Gdańsk.

Olejniczak, K. (2007), *Teoretyczne podstawy ewaluacji ex-post* (In:) *Ewaluacja ex-post. Teoria i praktyka badawcza*. A. Haber (Ed.), PARP, Warszawa.

Szara, K. (2014), *Czynniki warunkujące rozwój kapitału kreatywnego w regionach Ukrainy* (In:) *Studia wschodnie, Polityka – Gospodarka – Bezpieczeństwo*, Hudziakowani M., Zapałowski A. (Eds.), IG, Częstochowa.

Wallas, G. (1926), *The Art of Thought*, Jonathan Cape, London.

Wertheimer, M. (1968), *Productive Thinking*, Tavistock, London.

West, M.A. (2000), *Rozwijanie kreatywności wewnątrz organizacji*, PWN, Warszawa 2000.

Anna Pawłowska, PhD
Chair of Organization Theory
Faculty of Management
University of Warsaw

Lifelong learning motivation at the age of 45+. Some sources of limitations

Abstract: The chapter investigates how, and to what extent, the 45+ employees are eager to engage in further educational activity. On the basis of Vroom's expectancy theory, this study analyzes career development motivation. The research was conducted within a Polish-Danish project dedicated to labour market. The results support the increasing of lifelong learning programmes' effectiveness.

Keywords: lifelong learning, effectiveness, age, career development, employees

Introduction

Lifelong learning necessity is an obvious fact resulting from the specificity of labour market, driven by many various factors, such as advancements of modern technologies, knowledge-based economy, and very dynamic and highly competitive business environment. Thus, a compulsory implementation of programmes concerning lifelong learning is needed. The assumptions regarding lifelong learning should follow the recommendations of external labour market observers. Yet, the question arises whether the main labour market participants, i.e. employers and employees, have the same belief concerning their attitude to the necessity of lifelong learning as an obvious element of the functioning of labour market. For example, some previous research conducted among employers revealed that 20% of examined companies see no need for further training of employees aged 50+ (Kononowicz, Michałowska & Majewska, 2010). The employers themselves share stereotypes and therefore are not aware of the necessity of lifelong learning.

The chapter aims to deepen the understanding of the motivation of 45+ employees to engage in further educational activity. The analysis is focused on workers at the age of 45+, as they need to maintain their career activity. Learning is an obligatory condition for changing their market situation. Currently, in Poland only 28% of the 50+ population is professionally active, which is one of the worst results in the EU. Simultaneously, ageing people are the most passive group in terms of education, even though they face serious problems with finding a job

(Górniak, 2012). Among the unemployed aged 50–64, 88% are uneducated, while among the employed group at the same age the figure was 78%. Generally, the eagerness to participate in different courses and trainings decreases with age. In the group aged 50–64, only 9% declared their interest in further training (Górniak, 2012).

With the support of the European Union funds, a group of Polish researchers conducted a number of projects concerning the unwillingness for further education. However, it seems that the data obtained does not contribute to the improvement of lifelong learning programmes. Most of the studies are focused on statistic data and respondents' opinions. However, as R. René Schalk, B. van der Heijden, A. de Lange, M. van Veldhoven (2011) noticed, most investigations so far used a 'between-persons' approach focused on static differences, while the dynamic 'within-person' processes, mostly neglected, should be scrutinized.

The study concerning also individual attitudes towards ageing people's education tried to answer the following question: Do educational barriers really exist for people aged 45+? The results indicate that there is only one barrier, but a very powerful one – the existence of a very destructive stereotype that people at this age cannot learn and adjust to new situations (Urbaniak et al., 2008). It causes the negative attitude of employers towards this group of people: they neither want to hire employees at this age, nor to invest in them (as was already mentioned above). Moreover, common beliefs cause that the interested people start to believe in this stereotype and consider training as a very stressful test with predicted unsatisfactory results and therefore prefer to avoid it. Some claim that the source of this learning avoidance is placed in the mentality of ageing people (*Podnoszenie kwalifikacji...*, 2012).

However, considering it a stereotype or mental problem is too simplistic. It is therefore of great importance to conduct more detailed research on this issue. It would enable to optimize spending the funds for the career activity of people aged 50+, as their earlier effectiveness is unsatisfactory. For example, since 2009 Poland has spent for this purpose over 8 billion PLN from both domestic and EU funds. Unfortunately, according to Central Statistical Office data, these funds did not bring satisfactory effects, as 5% more than a year before people from this age group were still unemployed in February 2012 (90 thousand more than in 2009) (*Podnoszenie kwalifikacji...*, 2012). So, the motivation of ageing people in continuing education is worth further detailed examination in order to inform projects, activities and strategies enhancing employees' self-development.

Research approach and goal

The chapter attempts to answer the question concerning the reasons of the lack of educational activity among people aged 45+. It presents some considerations based on the results of the author's empirical research and analysis of the reports published in Poland within the recent years. The presented results have been obtained within the project "Adaptation of the Well-Box model as a tool to extend the activity of the 45+population on the Mazovian regional labour market", realized in partnerships with the Faculty of Management of the University of Warsaw and The Job Centre of Aarhus in Denmark in 2010–2012. The main goal of this project was to test if it is possible to adapt the Danish labour market instrument Wellbox to extend the activity of ageing people on the Mazovian labour market.

The conducted research was of both a qualitative and quantitative character. The first part embraced recruiting and selecting unemployed candidates for the project. It made use of the author's questionnaire created in order to diagnose the important factors concerning the possibility to estimate an effective implementation of the Danish tool on the Polish labour market. This chapter presents part of the results, which is related to educational activity of people aged 45+. In the quantitative part of the research, in turn, 52 unemployed people from two Mazovian regions took part, consisting of 19 men and 33 women. Moreover, also a Focus Group Interview (FGI) research among two unemployed groups was conducted. The first group included 15 people (4 men, 11 women), the second one included 14 people (11 men, 4 women) who passed the recruiting phase and were chosen to take part in the project. The conducted analysis and presented results regarding lifelong learning refer to the basic concepts of Vroom's work motivation theory.

The lifelong learning assumptions include the belief that learning is a way to get a job. A question may be asked to what extent people aged 45+ are motivated to work. One may refer to one of the motivational concepts, i.e. Vroom's expectancy theory. Due to this theory, a motivation to gain the objective is a function of the usefulness of the goal (how important and attractive it is for the individual) and subjective probability of gaining the objective (how much probability does the individual have in order to gain it). If one element stands for zero, then the whole motivation is zero.

Attractiveness of having a job

At the beginning, it may be considered whether work constitutes an aim in itself that is especially important for 45+ and whether they consider being employed as attractive. It may be considered at two different levels: socio-cultural and organizational that is expressed by employer-employee interactions.

Chudzikowski et al. (2009) put emphasis on cultural determinants in relation to career development. The importance of cultural values for the attitudes towards work of people 45+ in Denmark was emphasized by P. H. Jensen (2005). He indicates the percentage of population agreeing entirely or agreeing with the following statement: I would like to have a job, even though I don't need the money: Denmark 76.9%, France 52.7%, Germany 71.3%, Italy 51.7%, the Netherlands 51.7%, Spain 49.7%, England 55% (Jensen 2005). Although the Polish nation was not a subject of that research, some other studies indicate that being employed is not a value in the Polish culture and especially not in such a degree as it is in Denmark. Due to the Demoskop research in 1997, the situation was as follows: the less hours a day mature Poles worked, the more they felt as people with a high social position. In the research conducted a few years ago, no changes have been noticed in this regard – every second questioned pensioner claimed that nothing would make him work (Kononowicz, Michałowska & Majewska, 2010). According to the research results concerning Poles' attitudes, being employed in Poland is not a value in itself to such a degree as it is in Denmark, where the cultural meaning of work is very high. Therefore, having a job is not considered as a motivation factor encouraging further educational activity.

Participants of the Well-Box project who took part in the author's individual quantitative research gave answers with different results and indicated more optimistic approach than the Polish population at the general level. 67% of the examined people would rather rest with no need for earning money if they were granted a pension, but if they had a lot of money, they would rather work (79%). They claim that there are also other reasons why people want to work (71% 'yes' answers). However, it should be remembered that the questioned people were unemployed, but tried to find a job, and their answers may not be entirely considered as typical for most of Poles.

Yet, a positive work motivation among the ageing people seems to exists. The question is what factors in their work environment decrease their motivation instead of increasing it. According to R. René Schalk, B. van der Heijden, A. de Lange, M. van Veldhoven (2011), there is a need to better understand the intra-individual changes in employee-employer interactions over time (e.g. psychological contract relationships) in relation to work behaviour, enduring work ability and the related outcomes. Work and its attractiveness as a pursued value may be considered in the context of employee-employer interactions, as it may be a decisive factor whether an employee would like to work as long as possible.

Therefore, the participants of the Well-Box project were asked about typical employee-employer interactions. Quantitative and qualitative methods were used to diagnose this problem.

In quantitative research, respondents were asked two related questions. The first one embraced the importance of particular employers' behaviour towards them. The majority of respondents emphasized the importance of the employer's support and help, a clear definition of what to do, a friendly atmosphere at work, honesty towards employees, being able to understand employees and sensitive to different life situations. Moreover, the employer should treat employees as co-workers and be able to motivate and appreciate their work. Such appreciated criteria of positive employee-employer interactions were used in the second question, in which respondents were to indicate how employers usually behave towards employees. The percentage of 'yes' answers with reference to positive and negative interactions is presented in Table 1 below.

Table 1. Questioned people answers: How employers usually behave towards employees?

'Yes' answers describing the relations positively	Answers percentage	'Yes' answers describing the relations negatively	Answers percentage
Support and help	31%	Criticize and demand	69%
Clearly state what to do	42%	In a very general way state what to do and then are unsatisfied	58%
Are rather nice and friendly	65%	Are rather unfriendly and aggressive	35%
More often are honest towards employees	54%	More often are dishonest towards employees	46%
Treat employees as co-workers	46%	Treat employees with the great distance and superiority	54%
Encourage to work and appreciate efforts	44%	Do not see hard work, therefore rarely appreciate efforts	56%
Can understand employees and are sensitive to different life situations	35%	Are interested only in the company with no understanding towards employees	65%

'Yes' answers describing the relations positively	Answers percentage	'Yes' answers describing the relations negatively	Answers percentage
Are not very ambitious. Are satisfied with proper work.	58%	Are very demanding and give difficult tasks.	42%
Allow employees to present their own opinions concerning job tasks.	33%	Do not tolerate negative remarks. Employee is to listen and do tasks.	67%

Source: own elaboration.

Only in case of three factors, more than a half of the respondents gave answers indicating positive relations. They claimed that employers are rather nice and friendly, are not very ambitious and satisfied with well done work, and are more often honest towards employees. The rest of the interactions was perceived more negatively. The respondents stress especially the fact that, on the one hand, employers criticize and demand, but, on the other hand, do not tolerate any negative remarks. The employee is to listen and fulfil his or her tasks. Moreover, employers very generally state what should be done and then are unsatisfied, do not see the amount of work, rarely appreciate their employees, treat them with distance and superiority, are interested only in the business, and show a lack of understanding towards the employees. As may be concluded, the view of employee-employer interactions according to respondents is rather negative, and which is worst, highly differs from the 45+ expectations.

Unfortunately, the above phenomenon was proved by the qualitative research Focus Group Interview type. According to the respondents, dishonesty in the interactions of employee-employer type and among employees exists. Some exemplary utterances of respondents indicate it as follows:

1. "Now, every employee is quiet, cannot say anything… and if they don't and and tell the truth, they are fired, so you need to shut your mouth."
2. "Even if the co-workers see the whole situation, they rarely tell anything and if do so, they are reprimanded, better be quiet."

If employees do possess such a negative image of their work situation, it is difficult to expect that work will be such a value and attractive aim to pursue for them that they will eagerly take up any further training. They will rather limit their commitment and investment. Reaching the pension age and stopping professional activity is highly probable, unless an economic pressure appears.

According to Vroom's concept, the first factor significantly influencing work motivation is the attractiveness of the aim. It may be valued both from cultural perspective and work environment climate. However, within the researched context, this factor seems to be unattractive. This result supports the argument that it does not lead to the educational activity of ageing people.

Subjective probability of obtaining a job

Vroom's theory is the subjectively perceived probability of gaining expected results. So, adapted to the context of the research, the question is if the perceived chance of finding a job affects the motivation of the 45+ group to continue their education.

Interesting data were obtained in FGI research, in which the participants did not refer to their abilities and competencies as a necessary condition to get a job. Being asked a question what is needed in order to find a job, they answered – the employer. They were totally focused on external factors beyond their control. Such people do have an external locus of control. Age and appearance are for them the main barrier of getting a job. Among the necessary conditions to be employed are happiness, personal connections and cosmetic surgery. One of the respondents said:

"Experience of a 50 year old person and the age of a twenty year old."

The respondents' utterances revealed the assumption that finding a job depends on factors beyond their influence. Moreover, the respondents participating in the study claimed that they know a lot and are good employees with reference to their seniority and experience, and are astonished that employers do not want to employ them. According to them, employers share negative stereotypes and do not appreciate their potential and abilities, which is illustrated by the answer of one of the respondents:

> "Since I've been working in the same profession for 20 years, there is no need for me to be trained in order to start a new job."

In their answers, people aged 45+ do not concentrate on individual development but on the external factors and reveal negative attitudes towards them. In their opinion, qualifications and career experience, due to seniority and age, should make them better employees than younger people. Moreover, they claim that they know more and possess the courage to withstand the employer in contrast to young people. The response of a FGI research participant describes this attitude:

> "The employer does not appreciate anything. The young person may be controlled in an easier way… The young person does not have his own opinion. If an ageing person

is given a task, he considers it and eventually asks questions. The young one does not raise questions, he just acts. The ageing person asks either the employer or himself. If he notices that it is not proper, sometimes he is brave and says it, which is inconvenient for the employer. I think that employers consider themselves as masters, for they have the money and labour market behind themselves."

Respondents are convinced that employers are afraid of ageing people. Young employees are more subordinate and submissive. The conducted study and the obtained results support the previous findings (Kononowicz, Michałowska & Majewska, 2010). Paradoxically, although lacking employment, ageing people estimate their abilities at a very high level and do not see the need to further develop their competencies. This phenomenon is worth a more detailed analysis and has been already researched. A.H. de Lange, N.W. Van Yperen, B.I.J.M. Van der Heijden, P.M. Bal (2010) examined 450 pensioners of a temporary work agency in Denmark. Employing the achievement motivation concept, they analysed how ageing people define their aims with reference to the situations which require appropriate competencies at their work place. They found out that the aims of mastery-avoidance character are dominant at older age, which may be at the same time destructive for their work engagement and personal or social work importance. They are mainly focused on routine performance. Focusing on the avoidance aims, the authors of this research refer to the Socioemotional Selectivity theory of Carstensen (1995; cf. de Lange, Van Yperen, Van der Heijden, Bal, 2010). In light of this theory, individuals choose their aims depending on how they consider the future. Thinking about it, they may create limits or be open to it. It is very characteristic for the ageing people to make constraints, taking the perspective of time until death. It suggests that individuals are more eager to establish the avoidance aims with time. Taking the ageing people's perspective, one should have in mind the fact that training is a result of accepting the assumptions of more long-term aims. This may explain why it is both cognitively and emotionally difficult for 45+.

Perspective of career development theories

Debating the problem of 45+ education, the perspective of the career development concept is worth considering. According to the traditional approach, the dominant model assumed a stable career in one or two companies through a whole life, with some promotion to higher levels therein. Within this model, an employee acquired more experience and abilities with age, and the training system was under the employer's control. Such a perception of work and its rules is currently dominant among the 45+, and especially the 50+, as this kind of environment shaped their personal experiences.

According to Savickas, traditional concepts are useless for vocational behaviours analysis, as the world of business has changed in such a way that it cannot guarantee employees a stable career. Therefore, another career theory is proposed (Savickas & Porfeli, 2012). It assumes that man's development depends on adapting to the social environment and assumes a correlation between human and environmental factors. Regarding the career development, the individual has to adjust to the changing demands of employers and labour market (McMahon, Watson & Bimrose, 2012). One needs to be prepared for changing the employer a couple of times during his or her career activity, and even his or her profession. Within such a path, the individual must create the competencies portfolio that will assure his or her employment. Therefore, it is the individual's responsibility to acquire competencies relevant for the employer, to be trained in order to be later employed elsewher in case of being fired by the present employer. Therefore, career development researchers, e.g. Bańka (2006), indicate the training process externalization phenomenon. A situation, in which the employee takes responsibility for his or her competencies development, is new for people 50+. It is very difficult for them to find adequate behavioural models including lifelong learning. The modern labour market requires from them self-reliant thinking about themselves, their competencies, etc. In contrast, the traditional career model stated that with age the process of learning is less intensive. The lack of awareness of the rate of changes and the image of the world is the reason why people 45+ overestimate the importance of experience and job seniority, treating them as an advantage. These false assumptions concerning the features and rules of work environment seem to be another educational constraint of ageing people:

> "How did it used to be? I can describe it only in superlatives. Today, I cannot say the same about present employers".

> "I wish today's employers were the same as they used to be. In the past, you went to work with satisfaction".

It may be stated that people aged 45+ see themselves in the labour market context according to the traditional career development models. They demand a stable, long-term employment relationship. Their expectations and cognitive images are inadequate to the modern labour market realities that constitute the basis of lifelong learning programmes. Probably, at first they should be delivered with the knowledge concerning the rules of the modern labour market.

Conclusions

The conducted research and discussion allows us to better understand the causes of too low engagement of 45+ in lifelong learning. Although they have been presented from Poland's perspective, they seem to have a more universal character. The obtained results indicate some problems worth more detailed analysis and imply some suggestions for the improvement of the programmes. In particular, it would be beneficial for a more effective career activation of 45+ to implement programmes addressing directly the factors limiting their educational activity. Most of all, it is recommended to focus on helping them dealing with their own career development by developing such methods which would be adequate to their way of thinking. Education concerning the working rules in the modern company and the rules of the modern labour market are also of great importance. The element of influencing 45+ attitudes towards lifelong learning assumes also the employers' education, so that they are able to create a friendly work environment. At last, being employed should be promoted as a value in itself and one which is rooted in the national culture.

References

Bańka A. (2006), *Psychologiczne doradztwo karier*. Stowarzyszenie Psychologia i Architektura, Poznań.

Chudzikowski K., Demel B., Mayrhofer W., Briscoe J.P., Unite J., Bogicevic Milikic B., Hall D.T., Las Heras M., Shen Y., Zikic Y. (2009), Career transitions and their causes: A country-comparative perspective. *Journal of Occupational and Organizational Psychology*, 82, pp. 825–849.

Górniak J. (2012), *Bilans Kapitału ludzkiego*. PARP, Warszawa.

Heijden Van der B. I.J.M., Lange de A. H., Demeroutie E., Heijde Van der C. M. (2009), Age effects on the employability – career-success relationship. *Journal of Vocational Behavior*, 74, pp. 156–164.

Jensen P. H. (2005), Reversing the Trend from "Early" to "Late" Exit: Push, Pull and Jump Revisited in a Danish Context. *The Geneva Papers*, 30, pp. 656–673.

Kononowicz M., Michałowska J., Majewska A. (2010), *Osoby w wieku 50+ na mazowieckim rynku pracy*. PBS DGA Sp. z o.o. and Human Capital Business Sp. z o.o, Sopot.

Lange de A.H., Van Yperen N.W., Van der Heijden B. I.J.M, Bal P.M. (2010), Dominant achievement goals of older workers and their relationship with motivation-related outcomes. *Journal of Vocational Behavior*,77, pp. 118–125.

McMahon M., Watson M., Bimrose, J. (2012), Career adaptability: A qualitative understanding from the stories of older women. *Journal of Vocational Behavior*, 80, pp. 762–768.

Podnoszenie kwalifikacji zawodowych u osób 50+ http://www.egospodarka.pl/84993,Podnoszenie-kwalifikacji-zawodowych-u-osob-50,1,39,1.html (September 5, 2012.)

Ren´e Schalk R., Heijden van der B., Lange de A., Veldhoven van M. (2011), Long-term developments in individual work behaviour: patterns of stability and change. *Journal of Occupational and Organizational Psychology*, 84, pp. 215–227.

Savickas M.L., Porfeli E.J. (2012), Career Adapt-Abilities Scale: Construction, reliability, and measurement equivalence across 13 countries. *Journal of Vocational Behavior*, 80, pp. 661–673.

Urbaniak B., Samson H., Kołodziejczyk-Olczak I., Wieczorek I., Michno L. (2008), *Jak zachęcić pracowników po 45 r. ż. do dalszej edukacji. Rekomendacje praktyków*. Program Narodów Zjednoczonych ds. Rozwoju (UNDP), Warszawa.

Joanna Radomska, PhD
Wrocław University of Economics
Strategic Management Department

Risk associated with employee participation in the process of strategy implementation versus company size

Abstract: This chapter investigates the relationship between the risk associated with the employee's participation at the stage of strategy implementation and the size of the company. The results showed no correlation between them, so engaging employees at the stage of developing strategy and providing them with decision-making power is not burdened with a greater degree of risk in large companies.

Keywords: strategy implementation, strategic management, employee participation, risk, company size

Introduction

The problem of strategy implementation became a subject of research, which seems to be current and valid due to the fast pace and extent of changes in the business environment, which make it necessary not only to develop but also to implement an effective strategy. The less stable environment becomes, the more the need for the implementation of the strategy increases, resulting in the effectiveness and efficiency of current activities and supporting the achievement of success in the future. The rationale for undertaking research in the strategy implementation is the importance of implementation activities and the need to ensure consistency between the effects of the implemented projects, implementation programmes and their operational results. One of the methods is a broad communication of the implementation progress and a conscious inclusion of employees in this stage of the process of strategic management. The purpose of this article is to investigate the relationship between the risk appearing in the aspect of participation at the stage of strategy implementation and the size of the company. The research carried out aims to check whether there is a link between different types of risk associated with engaging employees in conscious measures to implement the strategy and the size of the organization.

Risk associated with employee participation in the process of strategy implementation – a literature overview

Risk is a threat which is difficult to predict, which may occur during the development and implementation of the strategy. It therefore appears necessary to take action to mitigate the impact of a number of existing risks, including that which is related to the inclusion of employees in the process of strategic management. Knowledge of this impact can also be used to generate benefits, because collecting and analysing information on the risks and taking them into account in decision-making can prepare for potentially adverse changes. In turn, the lack of risk analysis can lead to poor unfavourable decision for the organization or their improper implementation. The risk causes the threat of potential losses, but also can generate growth opportunities (Strategic Risk Management, 2000).

In the literature, we can find many definitions of risk relating to various aspects of the functioning of the organization. Many authors point to the aspect of risk, which is an essential part of the strategic management process (Noy & Ellis, 2003) and points to the existence of a link between risk and results achieved by the company (Jegers, 1991). From the perspective of strategic management, the most accurate seems to be the definition talking about the fact that it is a combination of factors, actions or activities causing a failure to achieve the objective (Kaczmarek, 2004). This set of elements can be considered as operational risk, which means the possibility of loss resulting from inadequate or failed internal processes, people and systems or from external events (Polish Financial Supervision Authority, 2004). This article is focused around one aspect of operational risk – the subject of employee participation and the related consequences. It seems that it is a key element from the point of view of the possibility to implement the strategy, and is still insufficiently explored in the existing literature. Many studies in fact refer to the psychological conditions related to strategic decision-making by managers (Herrmann & Datta, 2005) and the impact of emotions on strategic choices (Delgado-Garcia, de la Fuente-Sabate, de Quevedo-Puente, 2010). As part of the corporate risk management, the aspect referred to as the internal environment is pointed to, associated with the nature of the organization, work environment and employee attitudes (Committee of Sponsoring Organizations of the Treadway Commission, 2004). On the other hand, according to Speculand (2011), recognizing employees as stakeholders in the organization and taking implementation measures with their participation is one of the trends associated with the implementation of strategy, which is to be expected in the near future.

Risk associated with employee participation is significant, as it belongs to the so-called risk category I, and therefore avoidable risk. Included here are internal risk factors, linked inextricably with the activities carried out by the company, that can be controlled, eliminated and avoided, because their occurrence is not associated with any strategic benefits (Kaplan, 2012). In particular, management of this risk category includes measures for the development of organizational culture or establishing effective internal controls (Kaplan & Mikes, 2013). It is therefore related to the competence of managers, whose task is continuous analysis and interpretation of emerging threats (Kaplan, 2009) and adequate response to the chances observed (Berinato, 2004). And therefore having awareness of the natural character of the risk of participation is one of the elements of risk management in the organization (Buehler, Freeman & Hulme, 2008) including management of operational risk. It brings a number of benefits, including those related to increased confidence in the company by external stakeholders (Kersnar, 2009) and ensuring greater uniformity between the activities planned and those undertaken by employees (Simons, 2000).

In addition, as shown by Finkelstein and Hambrick, the degree of involvement of a wider group of employees has an impact on the decisions made by managers (Finkelstein, Hambrick, 1990). As indicated by the results of the studies carried out, this involvement may relate to all stakeholders, which, however, is associated with the threat of the emergence of conflicts caused by the functioning of the so-called interest groups (Miller, Hickson & Wilson, 2008).

Employee participation at the stage of strategy implementation is not a uniform process and can take different levels of advancement. The minimum level will mean, therefore, providing employees with information about the assumptions of the strategy and gaining the support of key employees from the point of view of opportunities for its implementation. A more advanced, but partial level, means an inclusion of employees in the implementation measures through participation in strategic projects, the use of implementation tools (such as BSC and strategy map), or involving them in the work when creating a schedule or budget. In this respect, however, the costs associated with the conclusion of a compromise that accompanies the implementation of conflicting interests should be taken into account. Then it is made up of alternatives optimal for all groups (Bourgeois & Eisenhart, 1989). Advanced level also includes the participation of employees at different levels in the work when developing the concept of development, and then its operationalisation. This applies to both managers as well as mid-level managers and other key employees to the implementation of the strategy. As indicated in the results of many authors, it is one of the success factors (Fahey & Randall, 1994); however, it requires the appropriate decision-making powers and delegation of authority (Brenes, Mena & Molina, 2008). It also involves costs which are generated by the need to devote more time and the risk of communication difficulties (Lines, 2007).

Each of the above-described levels of participation involves different types of risks, which are presented in the following table.

Table 1. Potential types of risk

Level of participation	Possible risks
Minimum	Employees do not know the strategy (insufficient or poorly maintained communication process) Key employees from the point of view of strategy implementation do not identify with it Employees do not see the link between current activities and strategy implementation (which in turn causes a decrease in the level of motivation)
Partial	Lack of acceptance of the strategy by employees Decrease in commitment Presence of interest groups
Advanced	Poor or insufficient exchange of information Too much information slowing down the decision-making process No incentive system linked to strategy implementation

Source: own work.

It is obvious that the types of risk described appear with varying frequency and severity in enterprises of all sizes, as they are related to both aspects of communication, and the elements characteristic of different sizes of organizations (e.g. presence of interest groups, which is more common in large enterprises). This is also effected by the current organizational structure, and thus the external and internal relations of managers at the highest level (Carpenter & Westphal, 2001) and the involvement of other key employees from the point of view of strategy implementation, whose roles differ between companies. It is also related to the possessed capital, including intellectual capital, whose impact on the realisation of the strategy remains undisputed (Rylander & Peppard, 2003). Another problem is the access to the resources necessary from the point of view of implementation of the development concept developed and the consequent limitations, often affecting smaller organizations (Mezger & Violani, 2011). It is therefore worth considering whether there is a link between the levels of participation specified and the related risk, and the size of the organization.

Hypotheses and research methods

To resolve the research issue described, the following hypotheses were formulated:

H1: There is a relationship between the risk associated with the minimum level of participation and the size of the company.

H2: There is a relationship between the risk associated with the partial level of participation and the size of the company.

H3: There is a relationship between the risk associated with the advanced level of participation and the size of the company.

The study was conducted using a direct questionnaire interview (PAPI), the quantitative survey used a method based on data collection in an open (explicit) and standardised manner. To ensure the highest representativeness, stratified random sampling, proportional in nature, was used, consisting of the selection from each strata of such number of elements which is in proportion to the number (share) of this strata across the population. On the basis of the calculations made for the level of significance defined at $α = 0.05$, the level of probable maximum fraction error within the main part of the study was determined at a level close to 5.4%.

The group of respondents included managers from 200 companies listed in two rankings. One hundred and one companies appeared in the *Polityka* magazine's list of 500 biggest Polish companies, where, besides sales revenue, the total revenues, gross and net profits as well as employment are also taken into account. Ninety-nine companies were included in the *Forbes Diamonds 2013* list. The Diamonds list includes companies which increase their value the fastest. As the study involved companies of different sizes, a division criterion of the number of employees was adopted, which allowed for analysing the results broken down into three sub-groups – 68 small companies (up to 49 employees), 63 medium-sized companies (50–249 employees) and 69 large companies (employing more than 250 people).

In order to study the relationship between two nominal variables, the Chi-square test of independence was used. Cramer's V coefficient was also determined for nominal variables, which measures the strength of the relationship between them.

Research results

Below are the research results for the hypotheses described earlier.

Table 2. Research results.

Hypothesis	Types of risk	Correlation coefficient	p	n
H1	communication process badly carried out	.332	.001	196
H1	lack of a sense of identity	.237	.087	196
H1	lack of connection of operational activities with strategy implementation	.262	.036	196
H2	lack of strategy acceptance	.252	.054	195
H2	decline in commitment	.185	.350	196
H2	interest groups	.261	.038	196
H3	poor exchange of information	.208	.206	196
H3	too large amount of information	.165	.503	195
H3	no incentive system	.177	.408	196

Source: own work.

The results shown in the table above indicate that only 3 correlations are statistically significant, which allows us to generalise the results for the entire population. Analysis of the results obtained for the first hypothesis showed that there is an average relationship between the risk of bad communication process and the size of the company. This translates into a lack of knowledge of strategy assumptions, and thus the formation of interference in the process of its implementation. As expected, this problem is characteristic of large organizations, which is confirmed by the results (52% of companies employing more than 250 people pointed to the existence of the barrier described). The risk in this case relates to engaging resources in the communication process, which is not efficient. This is related to the level of employment, as the greater number of employees intensifies the difficulties in the effective transfer of information. Interestingly, no correlation was found between the failure to identify with the development concept by key employees from the point of view of strategy implementation and the size of the company, which means that obtaining the support of the leaders of change is a matter just as important in organizations of different sizes. A relationship between the lack of knowledge of the employees about how their daily work contributes to the implementation of the strategy and the size of the company was also shown.

As before, this risk is most common in large organizations (48%) and is associated with the process of communication, which turns out to be ineffective. The results obtained allow for a positive verification of the hypothesis made.

In the case of the second hypothesis, no association was found between loss of involvement and a lack of acceptance of the strategy when employees were included in implementation measures, and the size of the company. The risk arising during the use of a dedicated tool is therefore similar in enterprises of all sizes and, therefore, should be considered in the same way by the decision makers in these organizations. However, a relationship between the presence of interest groups and the size of the company was found. It is a key aspect for as many as 55% of large companies. They are in fact exposed to a greater extent to the emergence of conflicts of power and lack of agreement during the allocation of funds and resources key to the implementation of strategy and strategic targets made. Therefore, the hypothesis made can only be partially accepted.

No correlation was found between the advanced level of participation and the size of the company. This means that engaging employees at the stage of developing strategy and providing them with decision-making power is not burdened with a greater degree of risk in large companies, and all organizations are exposed to it. It seems, therefore, that increasing the degree of participation is associated with the occurrence of risk of the same nature, regardless of the size of the company, its organizational structure or resources. It must therefore demonstrate an appropriate level of preparedness and strategy awareness to engage employees in the process of strategic management, particularly in its last stage – strategy implementation.

Conclusions

Engaging employees in the implementation of strategy is a difficult task, burdened with many types of risk. However, the benefits that are associated with this aspect are multidimensional and give a greater chance of implementation of the development concept developed, in line with the assumptions made. Size of the organization often determines not only the scope of participation, but also the manner of its introduction and, above all, the resources held. The results of the study showed, however, that this problem is mainly related to the minimum level of participation. It turned out that in this case the degree of risk is greater for large organizations. With the increase in the level of involvement of employees, the correlation with the size of the company decreases. The exception is the existence of interest groups that increase the risk for large companies. However, it may be concluded that the advanced participation of employees in the stage of strategy implementation requires the same level of attention of decision-makers, regardless of the size of the organization managed by them.

References

Berinato, S. (2004), Risk's rewards: are you on board with enterprise risk management? You had better be it's the future of how business will be run, *CIO*, Vol. 22, No. 3, pp. 1–10.

Bourgeois, L., Eisenhart, K. (1989), Strategic decision processes in high velocity environments: four cases in the microcomputer industry, *Management Science*, No. 34, pp. 816–835.

Brenes, E., Mena, M., Molina, G. (2008), Key success factors for strategy implementation in Latin America, *Journal of Business Research*, No. 61, p. 595.

Buehler, K., Freeman, A., Hulme, R. (2001), *Owning the Right Risks*, Harvard Business Review, September, p. 105.

Carpenter, M., Westphal, J. (2001), The Strategic Context of External Network Ties: Examining the Impact of Director Appointments on Board Involvement in Strategic Decision Making, *Academy of Management Journal*, Vol. 4, No. 4, pp. 655–660.

Delgado – Garcia, J., de la Fuente – Sabate, J., de Quevedo – Puente, E. (2010), Too Negative to Take Risks? The Effect of the CEO's Emotional Traits on Firm Risk, *British Journal of Management*, Vol. 21, pp. 313–326.

Committee of Sponsoring Organizations of the Treadway Commission. (2004), *Enterprise Risk Management – Integrated Framework*. Executive Summary, Wrzesień.

Fahey, L., Randall, R. (1994), *The Portable MBA in Strategy*, John Wiley & Sons, New York, p. 324.

Finkelstein, S., Hambrick, D. (1990), Top Management team tenure and organizational outcomes: the moderating role of managerial discretion, *Administrative Science Quarterly*, No. 35, p. 486.

Herrmann, P., Datta, D. (2005), Relationships between top management team characteristics and international diversification: an empirical investigation, *British Journal of Management*, No. 16, pp. 69–78.

Jegers, M. (1991). Prospect theory and risk-return relation: some Belgian evidence, *Academy of Management Journal*, No. 34, pp. 215–225.

Kaczmarek, T. (2006), *Ryzyko i zarządzanie ryzykiem. Ujęcie interdyscyplinarne*, Difin, Warsaw, pp. 52–53.

Kaplan, R. (2009), *Risk Management and the Strategy Execution System, Balanced Scorecard Report*, Harvard Business Publishing, November – December, pp. 7–8.

Kaplan, R. (2012), *Integrating Risk Management into the Strategy Execution System*, Harvard Business School, p. 13.

Kaplan, R., Mikes, A. (2013), Nowa koncepcja zarządzania ryzykiem, *Harvard Business Review Poland*, March, pp. 97–101.

Kersnar, J. (2009), Warning signs: why risk management is letting down companies and what to do about it, *CFO Europe Magazine*, February, pp. 25–30.

Lines, R. (2007), Using Power to Install Strategy: The Relationships between Expert power, Position Power, Influence Tactics and Implementation Success, *Journal of Change Management*, Vol. 7, No. 2, p. 152.

Mezger, S., Violani, M. (2011), Seven basic strategic missteps and how to avoid them, *Strategy & Leadership*, Vol. 39, No. 6, p. 21.

Miller, S., Hickson, D., Wilson, D. (2008), From strategy to action. Involvement and Influence in Top Level Decisions, Long Range Planning, No. 41, p. 607.

Noy, E., Ellis, S. (2003), Corporate risk strategy: does it vary across activities?, *European Management Journal*, No. 21, pp. 119–128.

Polish Financial Supervision Authority (2004), Recommendation M, Warsaw.

Rylander, A., Peppard, J. (2003), From implementing strategy to embodying strategy. Linking strategy, identity and intellectual capital, *Journal of Intellectual Capital*, Vol. 4, No. 3, pp. 326–327.

Simons, R. (2000), *Performance Measurement and Control Systems for Implementing Strategy*, Prentice-Hall, Upper Saddle River.

Speculand, R. (2011), Who murdered strategy?, *Strategic Direction*, Vol. 27, No. 9, p. 4.

Strategic Risk Management, Studies Research Paper WP07/00, Cambridge, United Kingdom.

Marta R. Jabłońska, PhD
Department of Computer Science in Economics
Institute of Applied Economics and Informatics
Faculty of Economics and Sociology
University of Łódź

Consumer needs and implementation of new technologies in the information society illustrated with the example of electric vehicle market

Abstract: Modern organizations are governed by the imperative of focusing their activities on clients. The chapter reviews the literature on organization-client relationships. On its basis, a research of customer attitudes towards electric vehicles was held. It identified the concerns and expectations of consumers, which could provide a basis for analysing and creating new customer needs on this market.

Keywords: electric vehicles, information society, customer relationship management, customer needs, new technologies

Introduction

Friedrich Nietzsche said that need is considered to be the cause of creation; in fact, it is often simply a result of what was created. This phrase, dating back to the nineteenth century, has a universal validity and can be applied to the current era – the information society. In everyday life, in which everyone notoriously encounters an excess of data – often completely unnecessary – a kind of information confusion arises easily. This phenomenon affects consumers and instils new thought, which in time will become a need and will lead to the purchase of goods or services. Excessive consumerism is also a determinant of the information society period. In numerous industries, such as new technologies, medicine, textile, automotive and telecommunications networks, marketing specialists are able to create new consumers needs and increase sales.

An example of those may be smartphones, whose popularity is still growing – in 2013 the number of such devices sold increased by 45% compared to the same period in the previous year (Komórkomania, 2014). At the same time, nearly one-third of smartphone users do not use the mobile Internet they offer (Life's

Good, 2014) and only two out of five admit that they took note of the features made possible by the device (Harris Interactive, 2014). Despite these statistics, consumers are willing to decide to buy a smartphone.

It is true that even the best technology will not become widespread without gaining social acceptance. This implies the actions of creating new customer needs described above. Managing the process of new product implementation is more likely to succeed when a dialogue with consumers is established. Understanding both the expectations and the potential concerns may allow for a better control of the process.

In this paper, the role of the analysis and creation of consumer needs in customer relationship management will be illustrated with an example of the automotive market branch, namely electric vehicles. For this purpose, a survey was conducted to gather information about the expectations and fears related to vehicles powered with electricity. The current share of electric vehicles in the total automotive fleet is negligible, due to their price and limited range. They are therefore a good example for the analysis of consumer needs that can support their further dissemination.

The paper is organized as follows. The second section was devoted to a review of the literature on consumer needs and their relationship with different organizations. Some considerations regarding the usage and dissemination of electric vehicles are the subject of the third section. In the following section, the results of the survey on the social attitudes towards electric vehicles were presented. The last part of the paper contains a summary, namely conclusions and further suggestions for a future paper.

CRM and customer needs and modern organizations

The fact that nowadays companies should be customer-oriented is usually taken for granted. A critique of this model usually appears in the case of focusing on consumer needs and assessing their impact on the organization and its functioning. In this section, a review of the current research directions on customer needs and customer-organization relationships will be presented.

Regardless of the fact that an organization wants to focus on creating new needs, satisfying the existing ones is important. Maintaining a dialogue with customers brings forth a specific feedback, which allows the detection of the organization's strong and weak points. The CRM system is often cited as the most popular method of gathering customer information (Arnett, 2005; Ertz, 2013; Luck, 2003; Ngai, 2005; Romano, 2002; Shaw, 2001; Wang, 1998). It provides comprehensive data management concerning sales, marketing and after-sales service, as well as the acquisition of knowledge about the products and the market.

What is more, the usage of CRM solutions such as Web mining makes it possible to acquire more knowledge about consumers (Xu, 2005) and their variable needs (Arnett, 2005), which, in turn, helps to predict emerging trends and gain customers' loyalty (Srinivasan, 1998).

The issue of loyalty is often related to the customer lifetime value (CLV). It is understood as an indication of the potential profits that a customer can bring to a company. CLV also provides tools to support customer relationship management (Audzeyeva, 2012; Berger, 1998; Donkers, 2007; Gupta, 2006; Jain, 2002; Malthouse, 2005). The acquired data can be used for customer segmentation according to the amount of expected revenues supplied to the organization. Moreover, CLV also provides the information necessary for decision-making on the modification of existing or implementation of new products (services) (Audzeyeva, 2012).

In the literature on customer needs and their relationship with organizations often actuated issue is the satisfaction of the clients, that can be a cornerstone of their loyalty (Chinomona, 2013; Liu, 2011; Rachmania, 2012). And so, research on customer loyalty in the mobile market are described, inter alia, in (Aydin, 2005; Deng, 2010; Katibi, 2002; Lai, 2009; Lee, 2010). The banking sector was presented in (Beerli, 2004; Hamadi, 2010) and the restaurant and fast-food in (Cheng, 2011; Han, 2009; Hwang 2010). Bugel, Verhoel and Buunk (2010) described determinants of customer engagement basing on 5 different sectors.

By focusing on a relatively narrow segment of the client-organization relations, that is, the analysis of needs, two different attitudes can be found in the literature. According to the first one, an important role that customer needs may play for organizations is highlighted (Fang, 2008; Franke 2006; Henard, 2001; Lieberman, 1998; Wang, 1998; Li, 1998). Such an attitude regarding the customer needs may indeed affect the introduction of completely new products and result in the advance of the competition. As a result, it even becomes possible to create new markets (Jaworski, 2000; Joshi, 2004; Stanko, 2013). Organizations aware of the current customer needs may also make a choice between releasing a product that meets them or attempt to create new client needs. On the other hand, there exists an attitude, according to which consumers are too short-sighted and their needs too variable to make any strategy on this basis. In addition, customers do not look ahead to the future and can even be unaware of their own needs. As a result, they are not able to identify the needs that would lead an organization to innovative solutions (Berthon, 1999; Campbell, 1999; Christensen, 1996). It should, however, be emphasized that the works describing such an attitude are dated from the end of the last century.

In addition to the CRM systems already described in the current section, it is also emphasized that in the information society an additional and important source of knowledge about the client, his attitude and needs, may be the Internet. It is therefore advisable to seek information posted by consumers on the Web, social networking sites, blogs or portals for collecting feedback and evaluation of products (services) (Stanko, 2013).

Irrespectively of some polemic appearing in the literature, the role of the analysis and creation of customer needs is widely described and seems to be an important issue. Maintaining a constant dialogue makes it possible to measure the levels of customer satisfaction with the purchased products (services) or to find their possible defects or deficiencies. It is also possible to obtain a loyal customer and create the conditions for the creation of the new needs, which may lead to subsequent purchases. A mechanism of the customer needs analysis and creation was presented in Figure 1. Through indirect (i.e. websites, advertising) or direct (usage) contact with the product, the customer shapes an attitude towards it. This process is influenced, among other things, by the intelligibility of information, confidence in manufacturer, the affordability of the product, possible fears or prejudices, the availability of substitutes or type of need that a product has to satisfy (existing, emerging). The management of the process focused on creating new needs requires a good contact with the customer and accurate communication that will strengthen the incentives and create the need. An insufficient dialogue can lead to searching for substitutes or weakening (and even disappearance) of the created needs. Organizations can use the process of creating new customer needs to improve sales of existing products or to increase the chances of the implementation of new ones.

Figure 1. Creating customer needs.

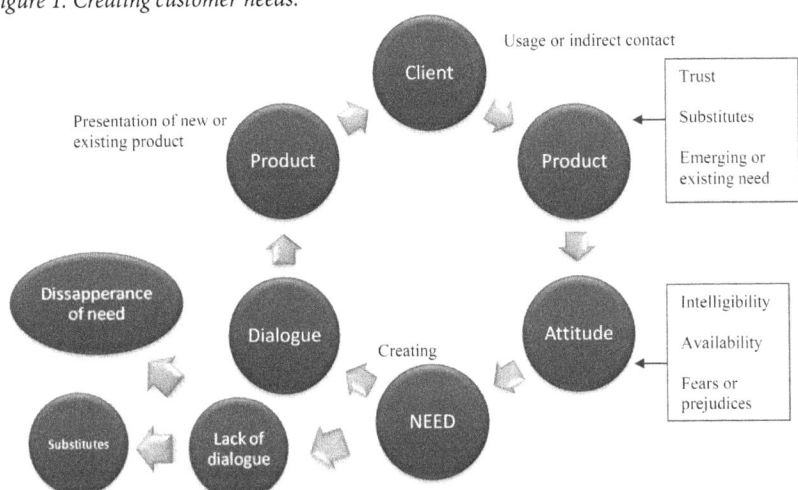

Source: own evaluation.

To sum up the considerations presented in this section, the management of customer needs is important for the increase of sales, implementation of new products and even creation of new markets. In the further part of the paper, these reflections will be referred to the electric vehicle market. Despite the optimistic comments (Hamid, 2013; Karfopoulos, 2013; Ortega-Vazquez, 2013; Propfe, 2011), these vehicles are not common and a way to their popularization is rather long. Any technology has reduced chances of spread, if it does not have social acceptance. For this reason, the author of the paper conducted a questionnaire survey to explain the concerns and expectations of consumers regarding vehicles powered with electricity. This may be a basis for an analysis of the existing needs and decision process of creating new ones.

Dissemination of electric vehicles

Electric vehicles are often considered as a remedy for the increased carbon dioxide emissions caused by transportation. The term electric reflects to both drives – the fully electric vehicles as well as various hybrid solutions that combine the usage of gasoline and electricity (Propfe, 2011).

Despite the optimistic predictions, according to which in the near future electric vehicles will dominate the transport by virtue of their advantages (the reduction of CO_2 emissions and fuel consumption, the support of power system operation), these vehicles are characterized by a number of limitations affecting the possibilities of dissemination (Hamid, 2013; Jabłońska, 2013; Karfopoulos, 2013; Ngai, 2005). Among the most important ones are: the impact on the power system and the additional supply of electricity required to charge the batteries, the insufficient number of stations and long charging time, the range and purchase costs (Liu, 2013; Lopes, 2011). What is also important is the fact that many combustion vehicles users do not possess any knowledge of their electrical counterparts. As a result, a need for purchasing an electric vehicle is still quite rare. Besides, a negative perception of electric vehicles is an effect of some myths.

The electric vehicles market is growing quite rapidly. New technologies increase range, reduce charging time and production costs (Yudovina, 2013). At the same time, more countries are introducing policies to support the implementation of electric vehicles and encourage consumers to buy them. However, these vehicles are still a niche product.

The dissemination of electric vehicles depends largely on the technical conditions. It is necessary to modernize the power system, namely by increasing the share of renewable energy in the energy-fuel balance (it is due to the fact that charging batteries with conventional energy results in CO_2 emissions similar to driving combustion-powered cars). Electric vehicles are likely to be loaded in the afternoon, which goes along with the period of peak electricity demand. This can lead to an overloading of the system and has a negative impact on the stability of its performance. For this reason, the popularization of electric vehicles is combined with the introduction of the so-called smart grid that can manage power system. Among other technical factors that should be mentioned, the attributes of electric vehicles and charging infrastructure such as increasing the number of stations and improving the performance of vehicles can contribute to an increase of their popularity.

Regardless of the technical conditions, to make electric vehicles widespread, an acceptance and willingness to buy them are required among consumers. An analysis of the existing fears and expectations can help to strengthen the emerging needs to purchase such a vehicle.

The study of social attitudes towards electric vehicles

The attempt to verify the attitudes (fears, expectations, willingness to buy or knowledge of the subject) was the main point of the research. This analysis can help in the process of determining the existing or creating new needs in this automotive market segment. In total, 250 respondents from Poland were involved.

The questionnaire was available in the electronic and paper forms. The statistics regarding demographic factors are presented in Table 1.[1]

Table 1. Respondents statistics.

	Number	Share %
Sex		
Female	91	38%
Male	147	62%
Residence		
Countryside	37	15%
City up to 25 000 inhabitants	22	9%
City from 25 000 up to 100 000 inhabitants	40	17%
City above 100 000 inhabitants	140	59%
Owner/user of combustion car		
Yes	159	67%
No	78	33%

Source: own evaluation.

In the first part of the research, the owners of electric vehicles were selected. Unfortunately, the share of these vehicles in the automotive market was also reflected in the study, because only 6 respondents declared their possession.

The second part of the study consisted of questions addressed to all respondents. The first of these concerned the advantages attributed to electric vehicles. Among the most popular answers were: modern technical solutions (22%), ecological (15%) and silent engine (19%). The distribution of responses is shown in Figure 2. In addition, respondents indicated: lower costs, plenty of power and high torque, brake energy recovery and high efficiency. The second question contained a catalog of disadvantages. The three most often mentioned were: price (27%), limited range (22%) and a small number of charging stations (23%). The distribution of responses is shown in Figure 3.

The third question was to define the purpose of possessing an electric vehicle in a household. 46% indicated that these vehicles are dedicated exclusively to

---

[1] The final report is available after contacting the author by e-mail: mjablonska@uni.lodz.pl.

short distances such as commuting, 41% for everyday use in the city and 13% considered that it will be proper in and out of the city.

Then a catalogue of factors that could affect the decision to purchase an electric vehicle was presented. Responses were as follows: the price of fuel (25%), the price of the vehicle (20%), incentives from the state (20%), reduction of CO_2 emissions (14%), modern solutions used in the vehicle (9%), performance (8%), appearance (4%). In addition, respondents emphasized the increasing the number of charging stations and the possibility of purchasing, if the vehicles would become common.

Figure 2. Advantages of electric vehicles.

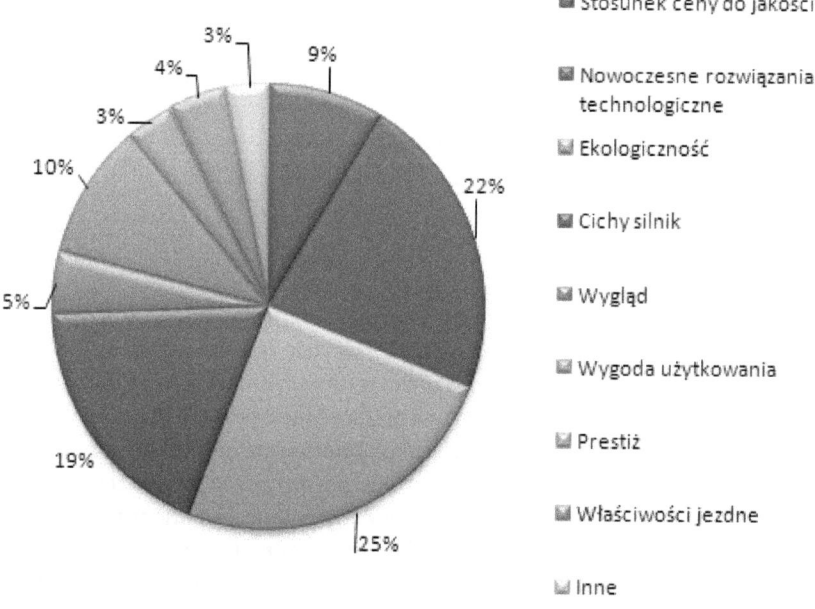

Source: own evaluation.

Figure 3. Disadvantages of electric vehicles.

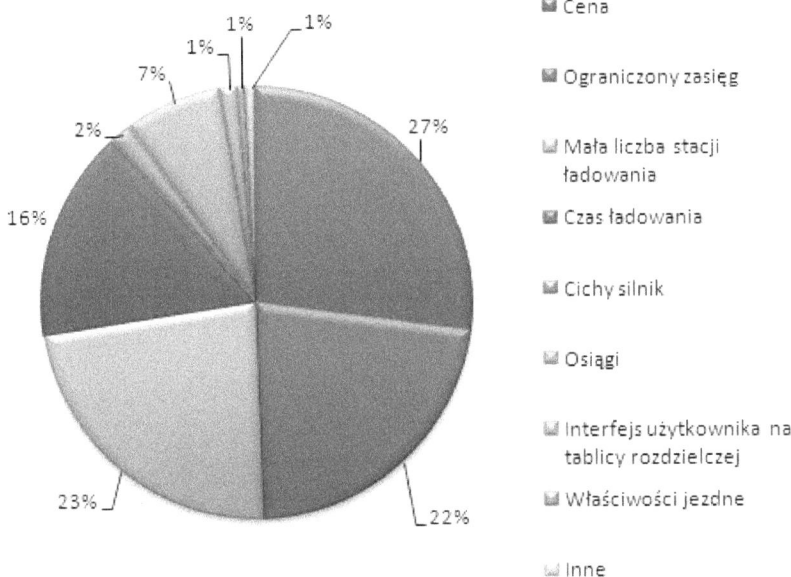

Source: own evaluation.

The price that respondents would be willing to pay for the electric vehicle was in the range up to 80.000 PLN (90%). 9% of respondents declared readiness to pay the amount of 80.000 to 100.000 PLN and only 1% was ready to pay more than 100.000 PLN. At the same time, 60% of respondents indicated that they do not wish that electric vehicle was the only means of transport in their household. With respect to the price of fuel, 52% of respondents indicated that they would be ready to buy an electric vehicle when the fuel price rises to 7 PLN/litre, 26% chose a value of 7 to 10 PLN and 22% above this amount.

In another question, respondents were presented with 13 models of electric vehicles (with the possibility of additional indications) and asked to select those which they had heard of. Among the most popular models were: Smart electric (16%), Tesla Roadster (11%), Chevrolet Volt (10%), Opel Ampera (9%), Nissan Leaf (9%).

In order to determine social attitudes more accurately, respondents were asked to identify their emotions caused by electric vehicles. Most of them declared that they are in favour of them, 1/8 feel the enthusiasm and every fourth – indifference.

Among the negative feelings, the majority of respondents chose distrust and fear and every twelfth reluctance. These results are illustrated in Figure 4.

Figure 4. Social attitudes towards electric vehicles.

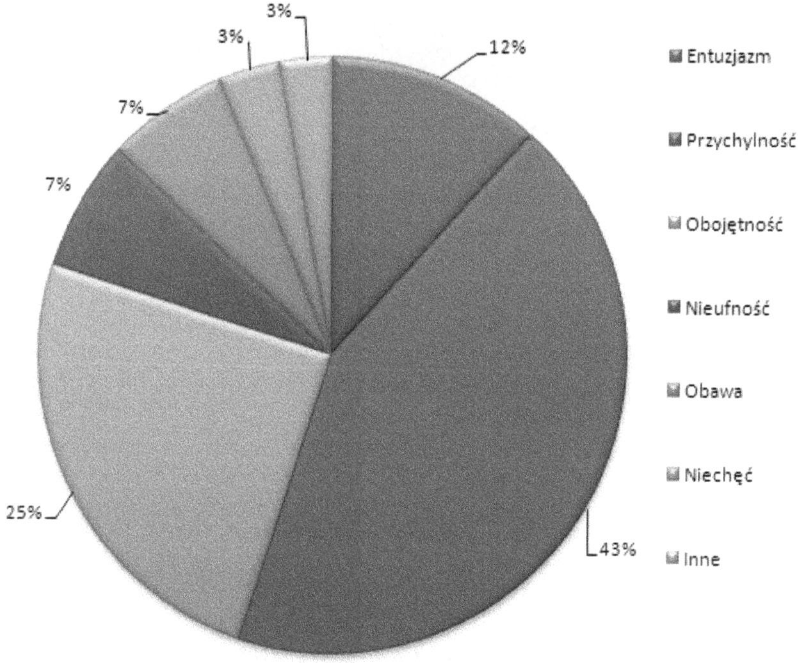

[Pie chart with values: 12%, 3%, 3%, 7%, 7%, 25%, 43%. Legend: Entuzjazm, Przychylność, Obojętność, Nieufność, Obawa, Niechęć, Inne]

Source: own evaluation.

Respondents were also asked about their concerns in the context of electric vehicles usage. More than ¼ of the respondents (28%) stated that these vehicles cannot be used for long journeys. A similar number of responses (26%) concerned too long charging time. Among others were: poor performance (17%), urban use (14%), not ecological (8%), concerns about safety (5%). Respondents also pointed to the cost of operation and replacing the battery, the price of vehicles and road service or problems with the use during winter. Only 13 of the 250 respondents said that they do not have any worries.

Conclusions

The increasing CO2 emission induces both the authorities and car manufacturers to seek alternatives to combustion vehicles. Although the idea of an electric car has been known for a long time, these vehicles are still not widely disseminated. As indicated in the third section dedicated to electric vehicles, their popularization is dependent on a number of technical conditions. However, with no or negligible demand for electric cars, the chance for their popularization decreases. The second section presented the role of the analysis and creation of customer needs in the information society. According to the author of this paper, a dialogue with potential electric vehicles users can help create new needs and thus increase the demand. The first step to achieve this cooperation can be the knowledge of the consumers' attitudes. Therefore, the research was conducted as described in the article.

The results indicate the possibility of creating the need to possess an electric vehicle. Nearly a half of the respondents refer to them favourably and enthusiastically, ¼ are indifferent, which also predicts the ability to change the position. Fears were caused largely due to a lack of knowledge or the limitations of a technical nature (as described in the third section) that manufacturers of electric vehicles have yet to overcome.

At the same time, the limitation stressed by the relatively low number of Polish respondents (N = 250) should be noted. In any further work concerning this subject, a similar survey at a larger scale should be carried out. For further discussion, also an indication of other aspects affecting the change in consumer attitudes that in the study were not included could be made. However, according to the author, the presented results may provide a starting point for further discussion and work on creating the consumers' needs in the electric vehicle market.

References

Arnett D.B., Badrinarayanan V. (2005), Enhancing Customer-needs-driven CRM Strategies: Core Selling, Teams, Knowledge management Competence, and Relationship Marketing Competence, *Journal of Personal Selling & Sales management*, Vol. XXV, No. 4, pp. 329–343.

Audzeyeva A., Summers B., Schenk-Hoppe K.R. (2012), Forecasting Customer Behaviour in a Multi-Service Financial Organization: A Profitability Perspective, *International Journal of Forecasting*, Vol. 28(2), pp. 507–518.

Aydin S., Özer G. (2005), The Analysis of Antecedents of Customer Loyalty in the Turkish Mobile Telecommunication Market, *European Journal of Marketing*, Vol. 39(8/9), pp. 910–925.

Beerli A., Martin J.D., Quintana A. (2004), A Model of Customer Loyalty in the Retail Banking Market, *European journal of Marketing*, Vol 38(1/2), pp. 253–275.

Berger P.D., Nasr N.I. (1998), Customer Lifetime Value: Marketing Models and Applications, *Journal of Interactive marketing*, Vol. 12, p. 17–29.

Berthon P., Hulbert J.M., Pitt L.F. (1999), Brand Management Prognostications, *Sloan Management Review*, Vol. 40(2), pp. 53–65.

Bugel M.S., Verhoel P.C., Buunk A.P. (2011), Customer Intimacy and Commitment to Relationships with Firms in Five Different Sectors: Preliminary Evidence, *Journal of retailing and Consumer Services*, Vol. 18, pp. 247–258.

Campbell A.J., Cooper R.G. (1999), Do Customer Partnerships Improve New Product Success rates?, *Industrial Marketing Management*, Vol. 28(5), pp. 507–519.

Cheng C.C., Chiu S.I., Hu H.Y., Chang Y.Y. (2011), A Study of Taiwan's package Tours, *Asia Pacific Journal of Tourism research*, Vol. 11 (10), pp. 97–116.

Chinomona R., Sandada M. (2013), Predictors of Customer Loyalty to Mobile Service Provider in South Africa, *International Business & Economics Research Journal*, Vol. 12, No.12, pp. 1631–1644.

Christensen C.M., Bower J.L (1996), Customer Power, Strategic Investment and the Failure of Leading Firms, *Strategic Management Journal*, Vol. 17(3), pp. 197–218.

Deng Z., Lu Y., Wei K., Zhang J. (2010), Understanding Customers Satisfaction and Loyalty: An Empirical Study of Mobile Messages in China, *International Journal of Information Management*, Vol. 30, pp. 289–300.

Donkers B., Verhoef P.C., Jong de M. (2007), Modeling CLV: A Test of Competing Models in the Insurance Industry, *Quantitative Marketing and Economics*, Vol. 5, pp. 163–190.

Ertz M., Graf R. (2013), *The Use of Web Mining for Existing Web Customer's Behavior Identification*, International Conference on Internet Studies, Hong Kong, China, available at: www.researchgate.net.

Fang E., Palmatier R., Evnas K. (2008), Influence of Customer participation on Creating and Sharing of New Product Value, *Journal of the Academy of marketing Science*, Vol. 36(3), pp. 322–336.

Franke N., Hippel von E., Schreier M. (2006), Finding Commercially Attractive User innovations: A Test of Lead-User Theory, *Journal of Product Innovation management*, Vol. 23(4), pp. 301–315.

Gupta S., Hassens D., Hardie B., Kahn W., Kumar V., Lin N., Ravishanker N. (2006), Modeling Customer Lifetime Value, *Journal of Service Research*, Vol. 9, pp. 139–155.

Hamadi C. (2010), *The Impact of Quality of Online Banking on Customer Commitment*, Communications of the IBIMA, pp. 1–8.

Hamid Q.R., Barria J.A. (2013), *Distributed Recharging Rate Control for Energy Demand Management of Electric Vehicles*, IEEE Transactions on power systems., Vol. 28, No. 3, pp. 2688–2699.

Han H., Ryu K. (2009), *The Roles of Physical Environment, Price Perception and Customer Satisfaction in Determining Customer Loyalty in the Restaurant Industry*, Journal of Hospitality & Tourism Research, Vol. 33, pp. 487–510.

Henard D.H., Szymanski D.M. (2001), Why Some New Products Are More Successful than Others, Journal of Marketing Research, Vol. 38(3), pp. 362–375.

Hwang J., Zhang J. (2010), Factors influencing Customer Satisfaction or Dissatisfaction in the Restaurant Business Using AnswerTree methodology, *Journal of Quality Assurance in Hospitality and Toruism*, Vol. 11, pp. 93–110.

Jabłońska M.R. (2013), *Social Aspects of Electric Vehicles Intrusion*, Proceedings in Virtual Multidisciplinary Conference, Zilina, Slovakia, pp. 171–174.

Jain D., Singh S.S. (2002), Customer Lifetime Value Research in Marketing: A Review and Future Directions, *Journal of Interactive marketing*, Vol. 16, pp. 34–46.

Jaworski B., Kohli A.K., Sahay A. (2000), Market-driven Versus Driving Market, *Journal of the Academy of Marketing Science*, Vol.28(1), pp. 45–54.

Joshi A.W., Sharma S. (2004), Customer Knowledge Development: Antecedents and Impact on New Product Performance, *Journal of Marketing*, Vol. 68(4), pp. 47–59.

Karfopoulos E.L., Hatziargyriou N.D. (2013), *A Multi-Agent System for Controlled Charging of a Large Population of Electric Vehicles*, IEEE Transactions on power systems, Vol. 28, No. 2, pp. 1196–1204.

Katibi A.A., Ismail H., Thyagarajan V. (2002), What Drives Customer Loyalty: An Analysis from the Telecommunications Industry, *Journal of Targeting. Management and Analysis for Marketing*, Vol. 11(1), pp. 34–44.

Lai F., Griffin M., Babin B.J. (2009), How Quality, Value, Image and Satisfaction Create Loyalty at a Chinese Telecom, *Journal of International Business Research*, Vol. 62, pp. 980–986.

Lee H.S. (2010), Factors Influencing Customer Loyalty of Mobile Service: Empirical Evidence from Koreans, *Journal of International Banking and Commerce*, Vol. 15(2), pp. 1–14.

Li T., Calantone R.J. (1998), The Impact of Market Knowledge Competence on New Product Advantage: Conceptualization and Empirical Examination, *Journal of marketing*, Vol. 62(4), pp. 13–29.

Lieberman M.B., Montgomery D.B. (1998), First-mover (dis)advantages: Retrospective and Link with the Resource-based View, *Strategic Management Journal*, Vol.19(12), pp. 1111–1125.

Liu C.T., Guo Y.M., Lee C.H. (2011), The Effects of Relationship Quality and Switching Barriers on Customer Loyalty, *International Journal of Information Management*, Vol. 31(2), pp. 71–79.

Liu H., Hu Z., Song Y., Lin J. (2013), *Decentralized Vehicle-To-Grid Control for Primary Frequency Regulation Considering Charging Demands*, IEEE Transactions on power systems, Vol. 28, No. 3, pp. 3480–3489.

Lopes J.A.P., Soares F.J., Almeida P.M.R. (2011), *Integration of Electric Vehicles in the Electric Power System*, Proceedings of the IEEE, Vol. 99, No. 1, pp. 168–183.

Luck D., Lancaster G. (2003), E-CRM: Customer Relationship Marketing In the Hotel Industry, *Managerial Auditing Journal*, Vol. 18(3), pp. 213–231.

Malthouse E.C., Blattberg R.C. (2005), Can We Predict Customer Lifetime Value?, *Journal of Interactive Marketing*, Vol. 19, pp. 2–16.

Ngai E.W.T. (2005), Customer Relationship management research (1992–2002). An Academic Literature Review and Classification, *Marketing intelligence & Planning*, Vol. 23(6), pp. 582–605.

Ortega-Vazquez M.A., Bouffard F., Silva V. (2013), *Electric Vehicle Aggregator/ System Operator Coordination for Charging Scheduling and Services Procurement*, IEEE Transactions on power systems, Vol. 28, No. 2, pp. 1806–1815.

Propfe B., Schmid S.A., Friedrich H.E. (2011), *Critical Paths and Sensitivities towards a Zero Emission Vehicle Fleet in Germany – A Scenario Based Approach*, Proceeding of Vehicle Power and Propulsion Conference 2011, available at: www.researchgate.net.

Rachmania I.N., Rakhmaniar M., Hani U., Wibisono D. (2012), ICOI 2012 Conference Proceeding, available at: www.researchgate.net.

Romano N.C., Fjermestad J. (2002), Electronic Commerce Customer Relationship Management: An Assessment of research, *International Journal of Electronic Commerce*, Vol. 2, pp. 61–113.

Ryu T., Shin S., Lim I-G, Oh K., Min D-K., Sun M., You H., Kim K-J., Yun M-H. (2003), *Development of Hierarchies of Customer Needs and Impressions for Passenger Car Interiors Based on a Survey of Car Reviews on Internet*, Proceedings of the XVth Triennial Congress of The International Ergonomics Association, Seoul, South Korea.

Shaw M.J., Subramaniam C., Tan G.W., Welge M.E. (2001), Knowledge Management and Data Mining for Marketing, *Decision Support Systems*, Vol.31, pp. 127–137.

Srinivasan S., Anderson R., Kishore P. (1998), Customer Loyalty in E-commerce: An Exploration of Its Antecedents and Consequences, *Journal of Retailing*, Vol. 78(1), pp. 41–50.

Stanko M.A., Bonner J.M. (2013), Projective Customer Competence: Projecting Future Customer Needs That Drive Innovation Performance, *Industrial Marketing Management*, Vol. 42(8), pp. 1255–1265.

Wahlberg O., Strandberg C., Sundberg H., Sandberg K.W. (2009), Trends, Topics and Under-researched Areas in CRM Research, *International Journal of Public Information Systems*, Vol. 3, pp. 191–208.

Wang R., Wen Q. (1998), Strategic Invasion in Markets with Switching Costs, *Journal of Economics & Management Strategy*, Vol. 7(4), pp. 521–549.

Xu M., Walton J. (2005), Gaining Customer Knowledge through Analytical CRM, *Industrial Management and Data Systems*, Vol. 105 (7), pp. 955–971.

Yudovina E., Michailidis G. (2013), *Socially Optimal Charging Strategies for Electric Vehicles*, (In:) Cornell University Library, http://arxiv.org/abs/1304.2329, accessed 2014.07.30.

www.harrisinteractive.com, accessed 2014.04.11.

www.komorkomania.pl, accessed 2014.04.11.

www.lifesgoodblog.pl, accessed 2014.04.11.

Monika Stachowiak-Kudła, PhD
Łazarski University

Evaluation as an instrument of higher education quality assurance

Abstract: This chapter characterizes evaluation as an instrument ensuring the quality of higher education. It analysed the internal quality assurance system requirements; identified basic functions and types of evaluation; and proposed criterions, indicators and methods of evaluation. The practice of evaluation has been scrutinuously outlined.

Keywords: quality assurance, evaluation, methodology, higher education, academic programmes

Introduction

One of the most important tasks of the government policy is the development of human capital. The level of this capital is mainly based on the quality of education, including higher education and scientific research (*National Development Strategy 2020...*, p. 100). According to the Ordinance of the Minister of Science and Higher Education of 5 October 2011 (§ 9 item 1 Point 9, § 9 item 4, § 10 Item 1), the government's activities in this area cover the specification and execution of the requirement of an internal quality assurance system, as one of the conditions for providing higher education services by universities. One of the instruments improving higher education quality assurances is quality evaluation.

The internal quality assurance system is a set of rules and procedures, activities and actions included in a formal catalogue, undertaken by a university's organs and internal stakeholders in order to improve the quality of education. The term internal system used in the above definition refers to the improvement of the quality of education inside a higher education institution, as opposed to the external quality assurance systems that are created in order to assess the guarantee of the improvement of quality assurance through an intervention of an institution superior to a university and entitled to undertake such an action (e.g. based on an Act) or a peer organization, e.g. experts from another university (Vlăsceanu, Grűnberg & Pãrlea, 2004, p. 74). It should be emphasized that the quality assurance should not be identified with two other terms prevailing in the management sciences, namely: quality management (QM) and the total quality management (TQM). In contrast to these two concepts, quality assurance does not apply to the activities of the entire university as an organization but only to the academic programmes/processes(Asif & Raouf, 2013, p. 2012).

The evaluation of an internal quality assurance system is used by basic organizational units of a university (faculties) in order to carry out an analysis of its own situation in the field of education. First of all, evaluation is an analytical and systematic process. Secondly, the analysis of a unit's situation performed on the appropriately collected and processed data. Thirdly, the evaluation outcomes are used in the process of decision-making and they are utilized in the teaching process (OECD, 2010, p. 21). The issue of evaluation as an instrument of higher education quality assurance has not been investigated intensively in the Polish literature on the subject, but it has been raised in the foreign one (Mizikaci, 2006, pp. 37-53; Karimyan, Naderi, Attaran & Salehi, 2012, pp. 719-728). Researchers tend to pay much more attention to the evaluation of educational systems (e.g. Mazurkiewicz, 2011, pp. 311-318) and the evaluation of projects financed by the European Union (eg. Turowski & Zawicki, 2007, pp. 29-58; Pylak, 2009).

The purpose of the article is to describe evaluation as an instrument of higher education quality assurance. To operationalize this purpose in the research context, the following specific objectives have been identified: the analysis of law regulations affecting the internal quality assurance system requirements for higher education, the identification of the basic functions and types of evaluation, the criteria, indicators and methods of evaluation, the number of universities pursuing an internal quality assurance system, and the identification of criteria and indicators that are actually used in the evaluation process (those available for higher education institutions).

In the research process, the following data collection methods have been used: an analysis of legislation and literature, an analysis of documentation published by universities concerning the quality assurance system and an analysis of information collected from the reports of institutional assessments carried out in 2012-2014 by the Polish Accreditation Committee. The choice of the Polish Accreditation Committee reports was determined by their completeness because they include a detailed description of the internal evaluation quality assurance system of all polish higher education institutions. They scope is much broader than any documents published by individual universities. The article refers to the internal quality assurance systems regulation covering the period 2012-2014.

Internal quality assurance system according to the Law on higher education

The Act of 27 July 2005 Law on higher education does not contain a legal definition of the internal quality assurance system. However, it refers to it in three articles. Firstly, Article 9, item 3, points 3 and 4 obliges the Higher Education Minister to specify by regulation the conditions for the assessment of academic programmes conducted by the Polish Accreditation Committee. They include the functioning of the quality assurance systems in the aspect of teaching outcomes, and an assessment of the development and improvement of the systems. Secondly, Article 66, item 2 specifying the duties of the rector of a public higher education institution to supervise the implementation and development of an internal system assuring the quality of education. Thirdly, Article 94b, item 1, point 3 states that the government budget shall designate an entity-specific grant for academic units of higher education institutions for the implementation of appropriate programmes enhancing the quality of teaching.

The concept of an internal quality assurance system was specified in two secondary regulations to the Law on higher education.

The provisions of the first of them – The Ordinance of the Minister of Science and Higher Education of 5 October 2011 on the conditions for studies in a specific field and at a particular level of education (§ 9 item 1 point 9, § 9 item 4, § 10 item 1, § 11) – classify the implementation of an internal quality assurance system taking into account the activities focused at improving educational programmes for the offered studies as one of the conditions for delivering studies. Moreover, they formulate requirements the system should meet: it should refer to all stages and aspects of the didactic process, especially including all forms of verification of learning outcomes achieved by a student with regard to knowledge, skills and social competence as well as the assessment made by students after each cycle of classes with regard to the fulfilment of didactic duties by an academic teacher, and the conclusions drawn from the university graduates' professional career monitoring. The provisions of the Ordinance thoroughly specify the obligation to verify learning outcomes: at the end of an academic year, the head of a basic academic unit, having asked the team of academic teachers that constitute the minimum staff of the specific field of study for their opinions, reports the assessment of learning outcomes to the council of the unit, which is then a basis for enhancing educational programmes.

On the other hand, The Ordinance of the Minister of Science and Higher Education of 29 September 2011 on the conditions of programmes and institutions assessment (§ 2 point 3 and § 7 point 2) classify the delivery of an internal quality assurance system as subject to the assessment by the Polish Accreditation Committee.

Article 53, item 1 of the Act of 27 July 2005 Law on higher education should be also discussed. Although it does not apply to an internal quality assurance system, it regulates: the form of the Polish Accreditation Committee's work, the criteria and the assessment procedures. In accordance with the Article, the Polish Accreditation Committee has adopted detailed criteria and procedures in its statute. For the further considerations, it is important to notice that Polish Accreditation Committee decided to use the notion of effectiveness as the assessing criterion of an internal quality assurance system. Irrespectively of the Polish Accreditation Committee's decisions, a higher education institution is expected to use the effectiveness criterion to assess its internal system and to use the evaluation outcomes for improving its quality assurance policy and to develop the education quality culture (*The criteria for assessing institutions...*).

Functions and types of evaluation

Evaluation is recognized in literature as a means of formulating judgments based on (qualitative or quantitative) indicators considered to be valid and reliable. With the use of these indicators, the actual results of an educational programme are compared with its anticipated results (Mizikaci, 2006, p. 41). It is also believed that even if evaluation concerns situations that are difficult to measure, it should remain reliable and it should be based on the data collected in a rigorous and objective way (Rossi, Freeman & Lipsey, 1999, pp. 84–85).

The key element is the continuous effort for an effective use of evaluation results. We have to make the recommendations developed in the evaluation process useful in building the new or improving the existing training programs.

Evaluation of an internal quality of assurance system fulfils four functions:

- a normative one – because it assesses the structure of activities and their value, collecting information necessary to plan activities in the future;
- a cognitive one – because it describes the mechanism of work of an internal quality assurance system in particular areas and assesses whether and how decisions taken within it are applied to solve potential problems;
- a reporting one – because the results of evaluation are communicated to the university authorities and its internal and external stakeholders; and
- an educational one – because it contains recommendations and conclusions for those who ordered the evaluation.

Evaluation of an internal quality assurance system can be adopted before the formal implementation of an internal quality assurance system (*ex-ante* evaluation), during it (on-going one) and after when the effects of system implementation become apparent (*ex-post* evaluation). An *ex-ante* evaluation helps to assure the relevance and coherence of the planned activities (Turowski & Zawicki, 2007, pp. 42–43). It focuses mainly on an analysis of the strengths and weaknesses of the system and the coherence with the university's mission and strategy. An on-going evaluation provides information that can improve the system management. Particularly the latter is worthy to be implemented in the new-developed quality assurance systems. An *ex-post* evaluation makes it possible to define real effects of an internal quality assurance system and makes it possible to draw conclusions for the future. It focuses more on the search for the current operational knowledge than on the processes of internal quality assurance. However, it is conducted too late to allow for the introduction of changes into the system but the conclusions and recommendations can improve the system in the next period. Therefore, this kind of evaluation should be considered as optimal. An *ex-post* evaluation should be carried out two to three years after the graduation of students completing the academic programme. The argument for this kind of evaluation is that some effects of the internal quality assurance system will reveal themselves when a student finishes the studies, e.g. those associated with employment.

Criteria, indicators and methods of evaluation

To evaluate an internal quality assurance system, one can use the following criteria: effectiveness, efficiency, relevance, impact and sustainability, and the criteria proposed by the European Network for Quality Assurance in Higher Education and the Polish Accreditation Committee, namely: transparency (*Standards and Guidelines...*, 2009) and complexity (*The criteria for assessing institutions...*).

Evaluation is essential especially for the determination of the system's effectiveness which measures how and to what extent the system has achieved its objectives (Blackmore, 2004, p. 134, Mizikaci, 2006, p. 41). It should be also pinpointed that evaluation of an internal quality assurance system is not easy. The most important difficulty is the measurement and assessment of the quality assurance system effectiveness, because it is relative (for example, it depends on how different players perceive and valuate the quality) and difficult to conduct. It is because the latter requires the measurement of long-term effects, proving the causal relationship between an internal quality assurance system and the improvement of quality.

The criterion of efficiency is defined as a proportion between the effects and the input (Turowski & Zawicki, 2007, p. 33). It can be used to evaluate an internal quality assurance system functioning in an academic organizational unit. In practice, however, such examinations seldom take place, mainly because it is difficult to establish the relation between effects and input with regard to e.g. the level of students' entrance or exit skills or the cost of achieving particular learning outcomes (Stachowiak-Kudła, 2013). The effectiveness of the whole university, however, can be examined (Sompolska-Rzechuła & Świtłyk, 2011, pp. 668–679; Rusielik, Świtłyk & Wilczyński, 2012, pp. 403–412). It is also necessary to emphasize that the literature on this topic points out that the maximization of outputs is not an objective of internal quality assurance systems (Sułkowski & Szewczyk, 2012, p. 246).

The relevance indicator makes it possible to answer the question whether, and to what extent, the aims of an internal quality assurance system meet the expectations of stakeholders. The real impact – concerning the results and influence of an internal quality assurance system – makes it possible to determine whether the system fulfils the expectations of stakeholders, whether it contributes to the solution of possible problems and whether its effects are beneficial for different groups of stakeholders. The sustainability of the internal quality assurance system effects makes it possible to assess whether the actions undertaken within the system are permanent and do not replicate the problems or anomalies which had occurred before. For the functioning of the system itself, the complexity is also essential. Firstly, it means that an internal quality assurance system covers all factors that influence the quality of education; secondly, it means the participation of all internal (academic teachers and students) as well as external stakeholders (employers and graduates) in the system. Finally, the transparency indicator makes it possible to assess whether institutions publish up to date, impartial and objective information about their programmes (*Standards and Guidelines…*).

In the recommended *ex-post* evaluation, the effectiveness criterion should be optimal. Assuming that the objective of academic programmes is to provide students with competence necessary for a satisfactory development of their long-term professional career on the domestic and international labour market, we can suggest some indicators helping to assess the effectiveness of the internal quality assurance system. The success rate seems to be the most important and includes the employment rate of graduates. Also the main stakeholder satisfaction index seems to be very important (Houston, 2007, pp. 3–17; Lo Franco & La Rosa, 2012, pp. 758; Chen, 2012, pp. 1288). Besides, it is worth using the academic satisfaction index – because employees are an enterprise' greatest asset and the satisfaction of customer is strictly linked with the satisfaction of employees (Nebeker et al., 2001, pp. 29–45, Chen, 2011, p. 86). Finally, one can suggest some additional indicators, like: an index of adequacy of examination and test questions (including the questions asked during the final/diploma examination) for the learning outcomes; an index of academic teachers who conduct classes directly connected with their own research (provided that the closer link with the research, the higher quality of the educational programmes); an index of academic teachers who have received a positive periodic assessment, etc.

The proposed indicators determine the type of sources of data necessary to conduct the evaluation. What is essential here are the documents of the educational and diploma awarding process as well as those regarding the monitoring of graduates' careers. With the adopted effectiveness criterion in mind, in the case of documents regarding the educational process, a special importance be attached to: the employers' involvement in the creation and improvement of educational programmes, checking and assessment of learning outcomes; the used methods and forms of learning outcomes verification; internship and placement; periodical assessment of academic teachers; students', graduates', employers' and academic teacher's opinions about the quality. The important sources of data are records of meetings: the faculty councils, committees for quality of education and representatives of employers.

The collection of data for the proposed indicators should rely on the examination of documents and social surveys (techniques of an individual in-depth interview and a focus group interview). Heuristic methods (techniques of expert panel, e.g. members of the committee for the quality of education) can be applied to analysis of gathered data. The examination of the internal quality assurance system should also apply more advanced methods like, for example, a quasi-experimental method (the study involving a group comparison).

Assessment of internal quality assurance systems in higher education institutions

In order to state whether the evaluation of an internal quality assurance system is common at the Polish universities, 50 institutional assessment reports, selected from 162 ones conducted by the Polish Accreditation Committee in 2012–2014 (http://www.pka.edu.pl/?q=pl/oceny, 2014), have been reviewed. The information on the subject issued by higher education institutions has also been analysed.

It should be also mentioned that the Polish Accreditation Committee seldom uses the term evaluation and prefers the broader concept of assessment. In the research, the activities undertaken by basic academic units in order to assess an internal quality assurance system are recognized as evaluation when the assessment is analytical and systematic (based on the appropriate, collected and processed information) and when its effects are used in the process of decision-making and applied to an individual in the process of learning.

According to the reports, 23 academic organizational units performed the evaluation of the internal quality assurance system. In the case of these units, the Polish Accreditation Committee indicated whether there are appropriate procedures allowing for the evaluation; secondly, whether they are functioning; thirdly, whether the evaluation outcomes are used for the improvement of the internal quality assurance system. 21 of these 23 units perform an evaluation systematically and two of them have done it for the first time.

The review of the reports shows that in five examined universities only some of their organizational units evaluated their internal quality assurance system. Nine reports contain information about methods used to evaluate an internal quality assurance system: eight reports mentioned a review of documents and one report – observation techniques and an expert panel. Based on the reviews of the reports, it is not possible to determine how many universities evaluate the effectiveness and/or efficiency of their internal quality assurance systems. Only in two reports, does the Polish Accreditation Committee consider the terms effectiveness and efficiency as different ones; in the rest of them, the terms are used interchangeably as synonyms. The review of the information available on the universities' websites suggests that none of them measures the level of efficiency of the internal quality assurance system, i.e. none of them examines the proportion of the input to the achieved outcomes/results. These universities focus on the effectiveness of the system and its complexity.

Evaluation uses quantitative indicators based on: inspection results, quality questionnaires, internal assessment of BA/MA theses, examination of students' scientific work, examination of the implementation of learning outcomes, examination of the availability of information about the implementation of the process of education etc., as well as qualitative indicators, e.g. students', graduates' and employers' opinions on the quality of education.

Conclusions

Summarizing the above postulates, there are strong arguments for the use of evaluation as an instrument for higher education quality assurance. The evaluation can answer the key questions: Did we accomplish what we have planned? What should we change? How should we do it? It is also a way to the implementation of higher education public governance based on the evidence (evidence based policy). This can further increase their operational effectiveness.

A little less than a half of academic organizational units in Poland assess their internal quality assurance systems. Such a poor result can be caused by the fact that the obligation to improve an internal quality assurance system was introduced into the Polish legal system recently. Despite the growing awareness and the need for quality assurance in educational programs, the methodology of evaluation of the internal quality assurance system is still underdeveloped. In this context, it seems very important to engage more researchers in the process of higher education quality assurance, even if they actually conduct studies on other fields of evaluation which are not directly linked with higher education. For example, it can include the scientist involved in the evaluation of: basic education quality or projects financed by the European Union. This is reasonable because the evaluation methodology remains similar regardless of the application area.

Last but not least, this recommendation is partly implied by the high concentration of Polish higher education institutions on the internal quality assurance system effectiveness, instead the efficiency of the system itself. In this context the criteria set by the Polish Accreditation Committee should be redefined and the verification of the universities' evaluation of the system should be abandoned because in fact this type of assessment is absent in the internal quality assurance. Finally, there is no reason to conduct the assessment of efficiency which is not carried out at the moment in higher education institutions.

References

Act of 27 July 2005 Law on higher education (uniform text: Journal of Laws 2012, No.0, item 572, as amended).

Asif M., Raouf A., (2013), Setting the course for quality assurance in higher education, *Quality and Quantity*, Vol. 45, pp. 2009–2024.

Blackmore, J. A., (2004), A critical evaluation of academic internal audit, *Quality Assurance in Education*, Vol. 12(3), pp. 128–135.

Chen, S.H., The establishment of a quality management system for the higher education industry, (2012), *Quality and Quantity*, Vol. 46, pp. 1279–1296.

Chen, S.H., A performance matrix for strategies to improve satisfaction among faculty members in higher education, (2011), *Quality and Quantity*, Vol. 45, pp. 75–89.

Clewes D., (2003), A Student-centred Conceptual Model of Service Quality in Higher Education, *Quality in Higher Education*, Vol. 9(1), pp. 69–85.

Houston D., (2007), TQM and higher education: a critical systems perspective on fitness for purpose, *Quality in Higher Education*, Vol. 13(1), pp. 3–17.

Karimyan H., Naderi E., Attaran M., Salehi K., (2012), Internal evaluation as an appropriate approach to improve higher education system; a case study, *Procedia – Social and Behavioral Sciences*, Vol. 69, pp. 719–728.

Lo Franco E., La Rosa S., (2012), Quality as an autonomous body of knowledge. An in-depth survey on Italian higher education, *Quality and Quantity*, Vol. 46, pp. 751–776.

Mazurkiewicz G., (2011), *Ewaluacja w nadzorze pedagogicznym. Model i system wartości*, (In:) *Ewaluacja w edukacji: koncepcje, metody, perspektywy*, B. Niemierko, M. K. Szmigel (Eds.), Kraków, pp. 311–318.

Mizikaci F., (2006), A systems approach to program evaluation model for quality in higher education, *Quality Assurance in Education*, Vol. 14(1), pp. 37–53.

National Development Strategy 2020, that is an annex to the Resolution No. 157 of the Council of Ministers of 25 September 2012 on the adoption of the National Development Strategy 2020 (MP of 2012, No. 0, item 882, as amended).

Nebeker D., Buss L., Werenfels P.D., Diallo H., Czekajewski A., Ferdman B., (2001), Airline station performance as a function of employee satisfaction, *Journal of Quality Management*, Vol. 6, pp. 29–45.

OECD, (2010), *Glossary of Key Terms in Evaluation and Results Based Management, DAC Working Party on Aid Evaluation*.

Ordinance of 29 September 2011 on the requirements for the assessment of programmes and institutions (Journal of Laws No. 207, item 1232).

Ordinance of the Minister of Science and Higher Education of 5 October 2011 on the conditions for delivering studies in a specific field and at a particular level of education (Journal of Laws No. 243, item 1445, as amended).

Pylak K., (2009), *Podręcznik ewaluacji projektów infrastrukturalnych. Czy Twój projekt przyniósł oczekiwane korzyści?* Warszawa.

Rossi P.H., Freeman H. E., Lipsey M.W., (1999), *Evaluation, A Systematic Approach*, Thousand Oaks.

Rusielik R., Świtłyk M., Wilczyński A., (2012), Efektywność publicznych uczelni technicznych w Polsce w latach 2007–2009, *Prace Naukowe Uniwersytetu Ekonomicznego we Wrocławiu*, Vol. 246, pp. 403–412.

Sompolska-Rzechuła A., Świtłyk M., (2011), Klasyfikacja uczelni wyższych w Polsce pod względem efektywności kształcenia – ujęcie dynamiczne, *Prace Naukowe Uniwersytetu Ekonomicznego we Wrocławiu*, Vol. 166, pp. 668–679.

Stachowiak-Kudła M., (2013), Dylematy oceny jakości kształcenia, Spektrum, Vol. 3(4), pp. XVIII–XXIII.

Standards and Guidelines for Quality Assurance in the European Higher Education Area (2009), European Association for Quality Assurance in Higher Education, Helsinki.

Sułkowski M., Szewczyk P. (2012), Wybór wskaźników dla oceny skuteczności wdrożonego systemu zarządzania jakością, *Zeszyty Naukowe Politechniki Śląskiej, series Organizacja i Zarządzanie*, Vol. 63a, pp. 245-258.

The criteria for assessing institutions that are an annex to the Statute of the Polish Accreditation Committee were adopted on 10 November 2011, http://pka.edu.pl/Dokumenty/Uchwaly/statut_final_10.11.2011.pdf.

Turowski B., Zawicki M., *Funkcje, etapy, metody i narzędzia ewaluacji*, (In:) *Ewaluacja funduszy strukturalnych – perspektywa regionalna*, Kraków 2007, pp. 29-58.

Vlăsceanu L., Grűnberg L., Pārlea D., (2004), Quality assurance and accreditation: A glossary of Basic terms and definitions, Bucarest.

hwww.pka.edu.pl/?q=pl/oceny.

Marzena Papiernik-Wojdera, PhD
Department of Business Analysis and Strategy
Faculty of Economics and Sociology
University of Łódź

Patterns of growth rates and real sales growth in Polish industrial companies in 2006–2012

Abstract: The objective of this paper is to assess the relation between real sales growth and selected concepts of company growth in the context of changes in the business cycle. The above objective was accomplished by conducting an empirical study using data illustrating the results of 61 industrial companies listed on the Warsaw Stock Exchange (WSE) in the years 2006–2012.

Keywords: sustainable growth rate, internal growth rate, corrected internal growth rate, real sales growth rate, company growth

Introduction

Effective and efficient sales management implies the need to take into consideration a wide spectrum of factors affecting the company's market and financial position. The paper discusses these issues in the context of selected concepts of company growth. The objective of this work is to assess the relation between real sales growth and sustainable growth rate, internal growth rate, and corrected internal growth rate. The above objective was accomplished by conducting an empirical study using data illustrating the results of selected joint-stock companies. The data were taken from the financial statements of 61 industrial companies listed on the Warsaw Stock Exchange (WSE) in the years 2006–2012. The analysis involved the real sales growth rate, the sustainable growth rate, the internal growth rate, and corrected internal growth rate in the context of changes in the total assets of the studied companies. The rates were calculated as arithmetic means for the industrial sector as a whole and for different industries individually, that is, the electronic and machine industry, the chemical industry, the wood and paper industry, the pharmaceutical industry, the light industry, the construction materials industry, the metallurgical industry, the automotive industry, the oil and gas industry, the food industry, and the plastic materials industry.

An analysis of changes in the above rates and in total assets was conducted based on data on the financial results of companies for the years 2006–2012, divided into three periods: 2006–2008, 2009, and 2010–2012, to elucidate the relationship between the studied rates and changes in the business cycle.

The analysis of trends in Polish gross domestic product (GDP) and sold production of industry in 2006–2012 is summarized by Figure 1.

Figure 1. Dynamics of GDP and sold production of industry in Poland in the years 2006–2012 (in %).

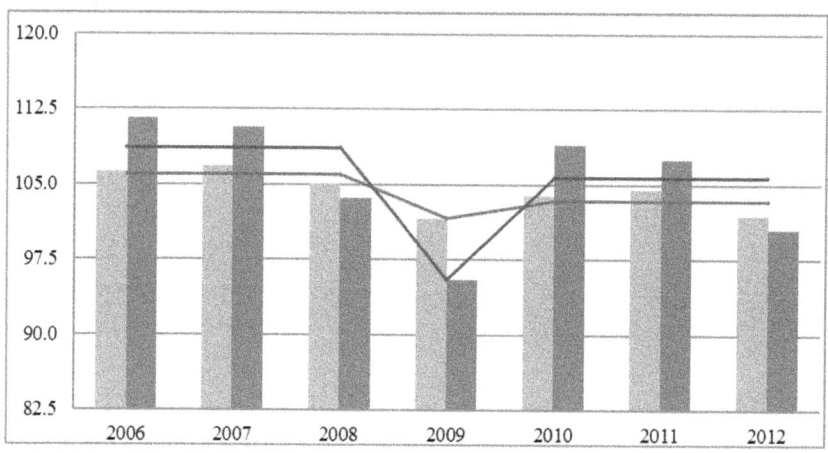

Source: Informacja o sytuacji…, 2014, p. 6; Roczniki Branżowe. Rocznik Statystyczny Przemysłu 2013, pp. 108–109.

Thus, the studied period exhibits three sub-periods that are very different in terms of the dynamics of economic activity in Poland. The study is based on data from the Emerging Markets Information Service, WSE InfoSpace, and publications of the Central Statistical Office (GUS).

Sustainable company growth, internal company growth, and corrected internal company growth

Company growth is perceived as a quantitative category reflecting an increase in the company's resources and leading to a greater scale of operations (Pierścionek, 1996). It is a gradual process of expansion of an economic entity. Company growth may be measured in different ways, with indicators typically referring to changes in sales, employment, assets, and market share. According to the economic literature, in situations where only one measure of growth is used, it should be based on sales (Delmar, Davidsson & Gartner, 2002), as they result from a demand for the company's products, reflect internal and external determinants and practices, and an increase in sales leads over time to the need of expanding the company's resources (assets, workforce) and affects its market position. Sales growth is a recommended measure of enterprise growth pursuant to the methodological standards of both Eurostat and the OECD. This measure of growth is used, amongst others, in business and entrepreneurship studies (See: OECD, 2007; GUS, 2013).

In formulating their strategic market goals, companies usually specify a certain level of sales volume, market share, or sales growth (Zoltners, Sinha & Lorimer, 2006). Attaining a certain level of sales has implications for the needs and possibilities of financing the activities enabling the achievement of the set goals. The sales growth rate that can be achieved under the financial policy pursued by a company may be determined using the sustainable growth rate (SGR) (Higgins, 1977; Kyd, 1986). Under this approach, the financial policy is analysed from the perspective of capital structure, dividend payout ratio, net profit margin, and total asset turnover. Sales growth depends on the potential of the company to self-finance (finance its growth with retained earnings without obtaining additional equity from the outside) while increasing debt within limits allowing for maintaining the company's capital structure.

In the studied companies, sustainable growth was determined using the following formula:

$$SGR = \frac{EAT_1}{E_0} \times er$$

where:

SGR– sustainable growth rate,
EAT_1– earnings after taxes at the end of fiscal year,
E_0– equity at the beginning of fiscal year,
er– retained earnings rate.

If a company does not intend to increase its debt or obtain additional equity from the outside, or if it has problems with accessing external capital, then sales growth may be financed only up to a level determined by retained earnings. Given the above, a useful notion in sales management is the internal growth rate (IGR)(Lee & Lee, 2006), expressed by the formula:

$$IGR = \frac{EAT_1}{TA_0} \times er$$

where:

IGR – internal growth rate,
TA_0 – total assets at the beginning of fiscal year,
other abbreviations as above.

In a situation in which a company does not pay out dividends (while not obtaining additional external capital), one can determine the corrected internal growth rate, which reflects the sales growth rate that can be financed with the generated net profit alone (without increasing debt or obtaining additional equity from external sources, and with retaining the entire net profit). Then, the retained earnings rate amounts to one, while the corrected internal growth rate equals return on assets, which is illustrated by the following formula:

$$IGR' = \frac{EAT_1}{TA_0}$$

where:

IGR' – corrected internal growth rate,
other abbreviations as above.

The above rates make it possible to evaluate different financial policy variants in companies differing in terms of size, resources, and sales financing methods. Thus, the question arises to what extent the above growth rates are consistent with the real sales growth rate in companies. Is it possible to find some patterns in these rates in economic practice, taking into consideration business cycle fluctuations? Does this imply differences in the effectiveness of asset use (asset intensity of sales)?

The above questions delineate the research area concerning patterns in company growth rates.

Sustainable, internal, and corrected internal growth rates and the real sales growth rate in industrial companies listed on the Warsaw Stock Exchange in the years 2006–2012

A comparison of the real sales growth rate and the sustainable growth rate in industrial companies listed on the Warsaw Stock Exchange (WSE) in the years

2006–2012 shows that in the analysed period average real sales growth was 4 times higher than sustainable sales growth (Figure 2). Therefore, industrial companies increased their sales at a rate that greatly exceeded the dynamics determined by the sustainable sales growth rate. This means that sales growth occurred under conditions of higher (rising) sales profitability and/or it was financed with additional external capital (with equity or with borrowed capital increasing leverage).

Given that in the years 2006–2012 the average annual sales growth rate in the industrial sector was 9.3% versus an average asset growth rate of 11.9%, the asset intensity of sales in that sector increased. Therefore, the effectiveness of asset use (as measured by total asset turnover) did not contribute to the creation of sales growth potential.

According to the internal growth rate, the studied companies could have increased their sales on average by 3.8% annually if they behaved pursuant to the principles of internal growth, that is, financed sales growth only with retained earnings and without obtaining any additional external capital. At the same time, it should be noted that the IGR was more than two times higher than the average IGR', which amounted to 1.6% annually. This means that in the studied period of time the potential for financing sales without dividend payouts was more than twice higher than that determined by the internal growth principle, that is, financing sales growth with retained earnings after dividend payouts and without obtaining additional external capital.

Figure 2. Average values of the real sales growth rate (RGR), sustainable growth rate (SGR), internal growth rate (IGR), corrected internal growth rate (IGR'), and growth rate of total assets (GRTA) in industrial companies listed on the WSE in the years 2006–2012 (in %).

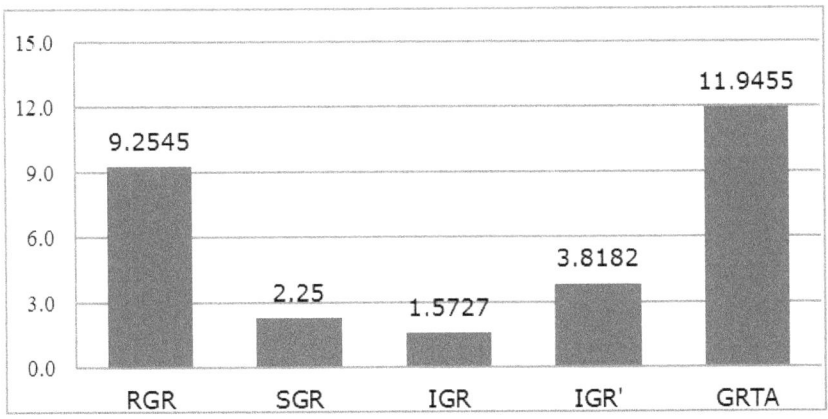

Source: Based on the financial statements of companies.

Considering the real sales growth rate, sustainable growth rate, internal growth rate, corrected internal growth rate, and the growth rate of total assets in the industrial sector across the three sub-periods analysed, one should note that in the year 2009, marked by the deepest economic slowdown, all the examined rates decreased (Figure 3). Real sales fell in the industrial sector by almost 13% on the previous year. Also the sustainable growth rate and the internal growth rates went into negative territory, with the corrected internal growth rate being the only positive indicator.

These data show that in the industrial sector sales growth could not be financed according to the principles of sustainable growth or internal growth. On the other hand, it was possible to finance a sales growth of 0.7% with net profits (without dividend payouts and without obtaining additional external capital), as evidenced by the corrected internal growth rate. It should also be remembered that in 2009 the industrial sector saw a considerable drop in total asset turnover as sales declined by 12.9% while total assets increased by 1.8%.

In contrast, both in the three years preceding and following 2009, all of the studied rates exhibited positive values (with the average sustainable, internal, and corrected internal growth rates being higher in 2006–2008 than in 2010–2012).

Figure 3. Average values of the real sales growth rate (RGR), sustainable growth rate (SGR), internal growth rate (IGR), corrected internal growth rate (IGR'), and growth rate of total assets (GRTA) in industrial companies listed on the WSE in the years 2006–2012 broken down into three sub-periods: 2006–2008, 2009, and 2010–2012 (in %).

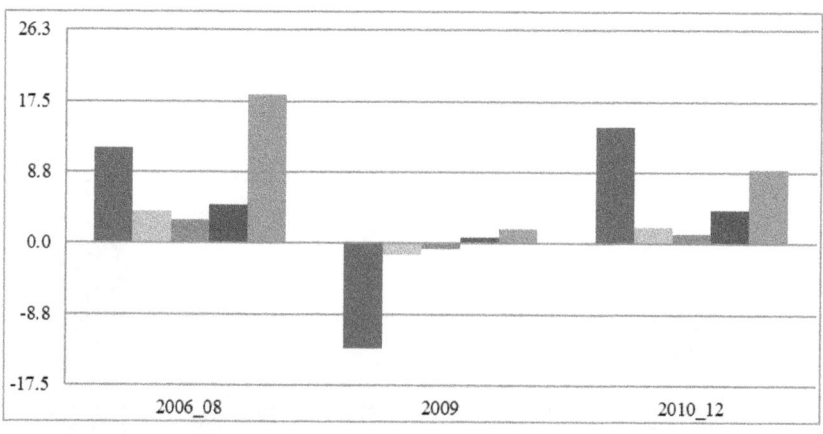

Source: Based on the financial statements of companies.

In the three years leading up to 2009, average sales growth dynamics in industrial companies amounted to 111.7%, while it was 114.3% in the years 2010–2012.

In the years 2006–2008, the average total asset dynamics was 118.3%, which was much higher than that in 2010–2012 (109.1%). This means that in the years 2006–2008, characterized by a relatively high economic growth (the average dynamics of annual GDP and sold production of industry was 106.0% and 108.6%, respectively), total asset turnover was lower than in the years 2010–2012, when economic growth slowed down considerably (the average dynamics of annual GDP and sold production of industry were 103.4% and 105.7%, respectively).

Analysing the average real sales growth rate and the growth rate of total assets, one may distinguish two major groups of industries (Tables 1 and 2).

Table 1. Average values of the real sales growth rate (RGR), sustainable growth rate (SGR), internal growth rate (IGR), corrected internal growth rate (IGR'), and growth rate of total assets (GRTA) in companies belonging to group I industries listed on the WSE in the years 2006–2012 (in %).

No.	Industry	RGR	SGR	IGR	IGR'	GRTA
1.	Metallurgical	23.8	6.1	4.3	6.3	14.3
2.	Automotive	3.6	4.2	2.6	5.5	3.3
3.	Oil and gas	20.1	7.1	4.0	5.2	13.4
4.	Food	10.2	1.5	1.4	4.2	7.9
5.	Mean values for group I	14.4	4.7	3.1	5.3	9.7

Source: Based on the financial statements of companies.

Table 2. Average values of the real sales growth rate (RGR), sustainable growth rate (SGR), internal growth rate (IGR), corrected internal growth rate (IGR'), and growth rate of total assets (GRTA) in companies belonging to group II industries listed on the WSE in the years 2006–2012 (in %).

No.	Industry	RGR	SGR	IGR	IGR'	GRTA
1.	Electronic and machine	9.0	5.5	3.5	6.8	11.1
2.	Chemical	5.0	5.6	3.2	10.2	14.7
3.	Wood and paper	4.8	7.2	1.4	3.5	8.7
4.	Pharmaceutical	14.1	-9.2	-2.2	-2.2	24.0
5.	Light	-1.7	-6.3	-2.8	-2.1	12.2
6.	Construction materials	5.1	-0.3	0.0	1.6	9.3
7	Plastic materials	7.8	3.3	1.9	3.0	12.5
8.	Mean values for group II	6.3	0.8	0.7	3.0	13.2

Source: Based on the financial statements of companies.

The first group consists of industries in which the average sales growth rate exceeds the growth rate of total assets (GRTA): in the studied period these are the metallurgical, automotive, oil and gas, and food industries. Thus, it can be concluded that in the years 2006–2012 this group exhibited increasing effectiveness of asset use. The second group is made up of those industries in which the real sales growth rate was lower than the growth rate of total assets (they revealed a decreasing total asset turnover); these are the electronic and machine, chemical, wood and paper, pharmaceutical, light, construction materials, and plastic materials industries.

In the first group of industries, the average annual sales growth rate amounted to 14.4%. while the growth rate of total assets was 9.7%. In turn, in the other group of industries, the average annual sales growth was more than two times lower (6.3%), at a considerably higher growth rate of total assets (13.2%).

Figure 4. *Average values of the real sales growth rate (RGR), sustainable growth rate (SGR), internal growth rate (IGR), corrected internal growth rate (IGR'), and growth rate of total assets (GRTA) in industrial companies listed on the WSE in the years 2006–2012 (in %) broken down into group I and group II industries.*

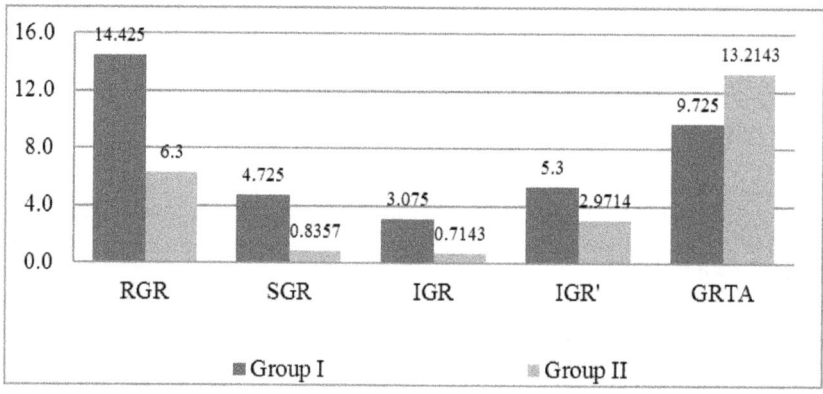

Source: *Based on the financial statements of companies.*

A comparison of the real sales growth rate, sustainable growth rate, internal growth rate, and corrected internal growth rate in the two groups of industries shows that these rates are higher in industries exhibiting more effective asset use, which means that they can finance sales growth more easily:

- under the conditions of sustainable growth (the sustainable growth rate amounted to 4.7% and 0.8% in groups I and II, respectively);
- by self-financing with retained earnings according to the internal growth concept (the internal growth rate amounted to 3.1% and 0.7% in groups I and II, respectively);
- by self-financing with all net profit kept as retained earnings according to the corrected internal growth concept (the corrected internal growth rate amounted to 5.3% and 3.0% in groups I and II, respectively).

Conclusions

Data analysis has led to the following findings concerning industrial companies listed on the Warsaw Stock Exchange in the years 2006–2012:

- the average annual sales growth rate (9.3%) was four times higher than the sustainable growth rate (2.3%);
- sales growth of up to 3.8% could be financed with entire net profit kept as retained earnings (according to the corrected internal growth rate concept);
- the potential to finance sales growth with retained earnings is two times lower if dividends are paid out (the average annual sales growth rate according to the internal growth rate concept amounted to 1.6%);
- the asset intensity of sales increased over time as total assets rose more quickly (on average 11.9% a year) than the real sales growth rate (9.3%).

The year 2009, marked by the deepest economic slowdown, saw the lowest values of the sustainable growth, internal growth, and corrected internal growth rates as well as the lowest sales dynamics and total asset dynamics in the entire studied period.

It should be noted that in the years 2006–2008 the sustainable growth rate, internal growth rate, corrected internal growth rate, and growth rate of total assets were higher than in the years 2010–2012, while sales growth was lower. In 2006–2008, sales grew on average by 11.7% annually, while total assets increased by 18.3%. In turn, the years 2010–2012 saw an average annual sales growth rate of 14.3% and a growth rate of total assets of 9.1%. This means that in the years leading up to the economic slowdown of 2009 the effectiveness of asset use was lower than that recorded during the period of slower economic growth after 2009. Thus, the years 2006–2008 seem to have been characterized by "investment optimism typical of periods of relatively intensive economic development, while the years 2010–2012 exhibited"investment restraint and greater dependence on existing production potential.

The various industries may be classified into two groups based on a comparison of the real sales growth rate and the growth rate of total assets. The first group consists of industries exhibiting increasing effectiveness of asset use (the metallurgical, automotive, oil and gas, and food industries). The other group contains industries characterized by decreasing effectiveness of the asset use (the electronic and machine, chemical, wood and paper, pharmaceutical, light, construction materials, and plastic materials industries).

As compared to the second group of industries, the first group shows higher real sales growth and higher average levels of sustainable, internal, and corrected internal growth rates with a lower growth rate of total assets. Thus, these industries use assets more effectively and have a greater potential of financing sales growth according to the principles of the analysed growth models.

References

Delmar, F., Davidsson, P., Gartner, W.B. (2002), Arriving at the High-growth Firm. *Journal of Business Venturing, 18*, pp. 189–216.

GUS (2013), *Wybrane wskaźniki przedsiębiorczości. Informacje bieżące* (wyniki wstępne: 29.11.2013). Download from: http://old.stat.gov.pl/cps/rde/xbcr/gus/PGWF_wybrane_wskazniki_przedsiebiorczosci_2013.pdf

Higgins, R.C. (1977), How Much Growth Can a Firm Afford?. *Financial Management, Vol.6*, pp. 7–16.

Kyd, Ch.W. (1986), How Fast Is Too Fast?. *Inc. Magazine, Vol. 8*, pp. 123–125.

Lee, Ch. F., Lee, A. C. (Ed.). (2006), *Encyclopaedia of Finance*. Springer Science and Business Media Inc, New York.

OECD (2007), *Eurostat-OECD Manual on Business Demography Statistics*. OECD and European Commission.

Pierścionek, Z. (1996), *Strategie rozwoju firmy*, Warszawa: Wydawnictwo Naukowe PWN.

Zoltners, A.A., Sinha, P., Lorimer, S.E. (2006), *How to Design and Implement Plans That Work: The Complete Guide to Sales Force Incentive Compensation*. AMACOM, New York.

New Horizons in Management Sciences

Edited by Lukasz Sulkowski

Vol. 1 Lukasz Sulkowski: Neodarwinism in Organization and Management. 2012.

Vol. 2 Lukasz Sulkowski: Epistemology of Management. 2013.

Vol. 3 Barbara Kożuch / Zbysław Dobrowolski: Creating Public Trust. An Organisational Perspective. 2014.

Vol. 4 Konrad Raczkowski / Lukasz Sulkowski (eds.): Tax Management and Tax Evasion. 2014.

Vol. 5 Michał Chmielecki / Łukasz Sułkowski: Metaphors in Management – Blend of Theory and Practice. 2017.

Vol. 6 Łukasz Sułkowski (ed.): Management and Culture of the University. 2017.

Vol. 7 Barbara Kożuch / Łukasz Sułkowski (eds.): Reflections about Contemporary Management. 2018.

www.peterlang.com

www.ingramcontent.com/pod-product-compliance
Ingram Content Group UK Ltd.
Pitfield, Milton Keynes, MK11 3LW, UK
UKHW022154230426
12049UKWH00004BA/96